REAL-TIME RENDERING

COMPUTER GRAPHICS
with CONTROL ENGINEERING

AUTOMATION AND CONTROL ENGINEERING
A Series of Reference Books and Textbooks

Series Editors

FRANK L. LEWIS, Ph.D.,
Fellow IEEE, Fellow IFAC
Professor
The Univeristy of Texas Research Institute
The University of Texas at Arlington

SHUZHI SAM GE, Ph.D.,
Fellow IEEE
Professor
Interactive Digital Media Institute
The National University of Singapore

PUBLISHED TITLES

Real-Time Rendering: Computer Graphics with Control Engineering,
Gabriyel Wong; Jianliang Wang

Anti-Disturbance Control for Systems with Multiple Disturbances,
Lei Guo; Songyin Cao

Tensor Product Model Transformation in Polytopic Model-Based Control,
Péter Baranyi; Yeung Yam; Péter Várlaki

Fundamentals in Modeling and Control of Mobile Manipulators, *Zhijun Li; Shuzhi Sam Ge*

Optimal and Robust Scheduling for Networked Control Systems, *Stefano Longo; Tingli Su; Guido Herrmann; Phil Barber*

Advances in Missile Guidance, Control, and Estimation, *S.N. Balakrishna; Antonios Tsourdos; B.A. White*

End to End Adaptive Congestion Control in TCP/IP Networks,
Christos N. Houmkozlis; George A Rovithakis

Robot Manipulator Control: Theory and Practice, *Frank L. Lewis; Darren M Dawson; Chaouki T. Abdallah*

Quantitative Process Control Theory, *Weidong Zhang*

Classical Feedback Control: With MATLAB® and Simulink®, Second Edition,
Boris Lurie; Paul Enright

Intelligent Diagnosis and Prognosis of Industrial Networked Systems,
Chee Khiang Pang; Frank L. Lewis; Tong Heng Lee; Zhao Yang Dong

Synchronization and Control of Multiagent Systems, *Dong Sun*

Subspace Learning of Neural Networks, *Jian Cheng; Zhang Yi; Jiliu Zhou*

Reliable Control and Filtering of Linear Systems with Adaptive Mechanisms,
Guang-Hong Yang; Dan Ye

**Reinforcement Learning and Dynamic Programming Using Function
Approximators,** *Lucian Busoniu; Robert Babuska; Bart De Schutter; Damien Ernst*

Modeling and Control of Vibration in Mechanical Systems, *Chunling Du; Lihua Xie*

Analysis and Synthesis of Fuzzy Control Systems: A Model-Based Approach,
Gang Feng

Lyapunov-Based Control of Robotic Systems, *Aman Behal; Warren Dixon; Darren M. Dawson; Bin Xian*

System Modeling and Control with Resource-Oriented Petri Nets,
MengChu Zhou; Naiqi Wu

Sliding Mode Control in Electro-Mechanical Systems, Second Edition,
Vadim Utkin; Juergen Guldner; Jingxin Shi

Autonomous Mobile Robots: Sensing, Control, Decision Making and Applications, *Shuzhi Sam Ge; Frank L. Lewis*

Linear Control Theory: Structure, Robustness, and Optimization,
Shankar P. Bhattacharyya; Aniruddha Datta; Lee H.Keel

Optimal Control: Weakly Coupled Systems and Applications, *Zoran Gajic*

Deterministic Learning Theory for Identification, Recognition, and Control,
Cong Wang; David J. Hill

Intelligent Systems: Modeling, Optimization, and Control, *Yung C. Shin; Myo-Taeg Lim; Dobrila Skataric; Wu-Chung Su; Vojislav Kecman*

REAL-TIME RENDERING

COMPUTER GRAPHICS
with CONTROL ENGINEERING

Gabriyel Wong

Jianliang Wang

CRC Press
Taylor & Francis Group
Boca Raton London New York

CRC Press is an imprint of the
Taylor & Francis Group, an **informa** business

Cover design by Gabriyel Wong

CRC Press
Taylor & Francis Group
6000 Broken Sound Parkway NW, Suite 300
Boca Raton, FL 33487-2742

© 2014 by Taylor & Francis Group, LLC
CRC Press is an imprint of Taylor & Francis Group, an Informa business

No claim to original U.S. Government works

ISBN-13: 978-1-4665-8359-7 (hbk)

Visit the Taylor & Francis Web site at
http://www.taylorandfrancis.com

and the CRC Press Web site at
http://www.crcpress.com

Especially for God, Crystal, Xavier, Xana, and Xaron, the love of my life.

G.W.

Could we with ink the ocean fill,
And were the skies of parchment made,
Were every stalk on earth a quill,
And every man a scribe by trade;
To write the love of God above
Would drain the ocean dry;
Nor could the scroll contain the whole,
Though stretched from sky to sky.

Frederick M. Lehman (1868–1953)

Contents

List of Figures

List of Tables

List of Abbreviations

3D   Three-dimensional (computer graphics)
ANFIS   Adaptive neuro-fuzzy inference system
ANN   Artificial neural network
ARX   Auto-regressive with exogenous (input or term)
BIBO   Bounded-input bounded-output
CAD   Computer-aided design
CAM   Computer-aided manufacturing
DTDNN   Distributed time-delay neural network
FPS   Frames per second (also known as frame rate)
GPU   Graphical processing unit
GUI   Graphical user interface
LAN   Local area network
LoD   Level of detail
MISO   Multiple-input–single-output
MLP   Multi-layer perception
N4SID   N4 subspace identification method
PID   Proportional, integral, derivative (control)
QoS   Quality of service
SISO   Single-input–single-output
SNPID   Single neuron PID
TCP   Transmission control protocol

Preface

Interactive computer graphics is a mature field of study. In fewer than 15 years, the improvements in speed and realism of computer-generated graphics from even consumer grade computers have been phenomenal. There is no lack of evidence to substantiate this statement as we observe the ever-increasing number of cutting-edge interactive applications such as computer games, virtual prototyping, and visualisation software. However, real-time computer graphics applications are often oriented toward meeting a particular set of goals without consideration of some form of global optimisation. A number of years ago, through real-life encounters in large-scale system implementation, the idea of convoluting computer graphics rendering with control theory was born.

From a larger perspective, computer graphics rendering is akin to any other process that runs on a computer. In recent years, researchers found that the increasing inclination to employ control engineering techniques in computer-related processes is not so much a matter of computer control (using a computer as a controller) as controlling the processes within a computer. Examples of such implementation are discussed in the vast array of research literature about server performance, network traffic control, and adaptive software with defined quality-of-service metrics. We believe the trend is no coincidence; it represents wide acceptance of benefits from integrating control theory with computer processes.

Our motivation for this work is simple. First, we want to provide a fundamental analysis of interactive computer graphics rendering from a systems perspective. Second, we want to establish a framework that facilitates interactive computer graphics rendering in an environment providing optimal utilisation of resources and good responses to rendering load changes. These goals can be accomplished through the adoption of digital signal processing, system identification and control engineering techniques that we believe will draw the interest of researchers and practitioners in the computer graphics-related fields.

While classical control demands meticulous evaluation of numerous criteria, the goals of our control system described in this book focus on tracking user-defined performance objectives while providing good transient responses so that changes arising from rendering load control will not lead to abrupt changes in visual displays. Furthermore, unlike physical systems utilised in aircraft, motors, and chemical mixers in which a failure of a control mechanism may lead to a catastrophic outcome, interactive computer graphics rendering is generally fail-safe.

In the course of this work, the computer graphics rendering process is modelled from a data-driven and black-box approach. We have shown the possibilities of various input–output configurations in a system model setting. While some may argue that the rendering process is too complex to be modelled by a few variables, we hope the reader can appreciate that the modelling technique in this book is in fact not congruent to this argument, but rather a systematic approach because the derived system models are substantiated with measured data.

Finally, it is our sincere hope that this work can further stimulate cross-disciplinary research and provide a premise upon which more interesting modelling and control techniques for real-time computer graphics may be developed.

MATLAB® is a registered trademark of The MathWorks, Inc. For product information, please contact

The MathWorks, Inc.
3 Apple Hill Drive
Natick, MA 01760-2098 USA
Tel: 508-647-7000
Fax: 508-647-7001
E-mail: info@mathworks.com
Web: www.mathworks.com

Acknowledgements

Words are just inadequate to express my gratitude toward Professor Wang Jianliang, who is more than just my supervisor, he is a mentor and friend for many years. Through him, I have learnt to appreciate the beauty of control theory. More importantly, his enduring encouragement and support have left me with a deep appreciation of him as a true educator. If there is one conversation I would choose to remember for life, it would be when he distilled the spirit of academic research as a pursuit of excellence and challenge.

This book would not have been possible without the support of many, especially my family. My heartfelt appreciation goes to my parents who did not have an extensive education yet believed wholeheartedly in the value of continual education, to the extent of making sacrifices for me in so many ways. To me, there is no closer personification of selfless love than this. My wife Crystal, the gem of my life, has been most instrumental in this endeavor. I wish to thank her for carrying the burden on the home front and being such a dedicated partner. She is a godsend whom I can never do without in every season of my life. The credit and fruit of this labor belong to all of them.

Last but certainly not least, I thank God for this journey of molding and growth. I thank Him for all the people who have made a difference in my life through this work and every step which He has hand-held me. To complete writing this book is a task that requires unimaginable perseverance and strength which He has so graciously given to me.

Gabriyel Wong

Summary

The value of interactive computer graphics is underscored by myriad applications in many domains of our lives. Consumers today can expect extremely realistic imagery generated in real time from commodity graphics hardware in applications such as virtual prototyping, computer games, and scientific visualisation. However, the constant and increasing demands for fidelity coupled with hardware architecture advancement pose many challenges to researchers and developers as they endeavour to find optimal solutions to accommodate speed of rendering and quality in interactive applications with real-time computer graphics rendering. The qualitative requirement of such applications, apart from the subjective perception of the displayed imagery, is the response time of a system based on user input. In other words, the requirement translates to the speed at which the machine can produce a rendered image according to the input provided by the person in the loop of the feedback system.

Earlier research attempted to address the frame latency problem by providing mathematical models of the rendering process. The models were often primitive because they were derived from coarse approximation or depended on specific application level data structures. Most approaches are based on heuristics and algorithms and are largely dependent on a specific type of application corresponding to the research. A major shortcoming of such techniques lies in the non-guarantee of performance.

From a systems perspective, the rendering process is modelled from an open-loop approach underpinned by constraints and estimations of the constituents of the rendering process. As a result, the output often fluctuates within an acceptable performance range. Furthermore, many such techniques rely on specific hardware or they may require unfriendly implementation on current computer graphics hardware. The advent of more sophisticated consumer graphics hardware in recent years has caused the rendering pipeline to be used in a far more complex manner to achieve ultra-realistic visual effects. Consequently, adapting models into applications becomes progressively more challenging as hardware and software technologies continue to evolve.

We can see from this background the exciting opportunities for the introduction of modelling and control principles into existing computer graphics systems. Our research focused on a systematic approach to realising a framework for modelling and control of real-time computer rendering in two stages:

1. Investigation, analysis, and implementation of a data-driven system identification process for real time rendering
2. Structured analysis of the derived model for the selection and design of a suitable control strategy

The first part of this book focuses on the modelling aspects of real-time rendering. Based on the dynamic natures of the possible and myriad variations of render states, polygon streams, and the non-linearity of the rendering process, we propose a data-driven modelling approach that accurately represents the system behaviour of this process from two angles: (1) the larger operating range where non-linearity exists and (2) the piecewise linear operating range. We propose two techniques for tackling the modelling challenge: (1) using a feed-forward time delay neural network derived from experimental data and (2) fuzzy modelling.

We demonstrate that both techniques can yield very accurate results in comparison with actual measured data. In addition, we compare the estimated outputs of our models with other mathematical estimation methods to show that the models derived from our approach yield better results than mathematical estimations. Starting with single-input–single-output (SISO) system models, we extend our work to investigate the validity of multiple-input–single output (MISO) systems as well.

The second part of this book focuses on the design of a control strategy based on the process nature investigated in the earlier chapters. The benefits of applying control theory in the context of a computer graphics system are explained and the relative advantages of the theory over the performances of existing heuristics and algorithms (open-loop estimations of rendering) are highlighted.

Our research proposed two controller designs to achieve stable output with accurate tracking: (1) proportional, integral, derivative (PID) control and (2) neural and fuzzy control. We investigated control system implementation in both local and distributed configurations.

In the local configuration, the rendering process ("plant") and controller reside in the same computer. In the distributed configuration, the controller runs on a computer different from the one used for rendering. The control activities and plant feedback are communicated between the computers via a network link. Despite network latency, this configuration allows flexible usage of system-wide resources in an integrated environment. The approach will be especially useful if elaborate controller designs adopted in the future result in the introduction of heavy computational loads into overall systems.

Authors

Gabriyel Wong is an entrepreneur, innovator, and author with extensive experience spanning leadership, managerial, and consulting roles in technology businesses. He currently works for one of Europe's largest private equity businesses in e-commerce and heads product performance and strategy activities in Southeast Asia.

Wong was the co-founder of XPEGIA, a Singapore-based start-up specialising in interactive media solutions for the advertisement and education markets. Before that, he was the R&D Director at EON Reality, a global leader in virtual reality technology based in the United States; he spearheaded the company's research and development. Before joining EON Reality, Wong was the founding director of gameLAB, the first research laboratory in Singapore to focus on computer game design and technology. He was a faculty member at Singapore's Nanyang Technological University (NTU) and lectured in both undergraduate and post-graduate programs.

He started his career as a technical lead at Singapore Technologies, one of Asia's largest engineering conglomerates and led the pioneering work on advanced computer graphics technology for defense applications.

Wong has published papers and spoken at conferences around the world and secured public and commercial funding for patenting his inventions. He earned B Eng and M Eng degrees in 2000 and 2012 from NTU and will be earning his PhD in 2013.

Jianliang Wang, PhD, earned a BE in electrical engineering from Beijing Institute of Technology in China in 1982 and pursued MSE (1985) and PhD (1988) degrees in electrical engineering from The Johns Hopkins University in the United States.

From 1988 to 1990, Dr Wang was a lecturer in the Department of Automatic Control at Beijing University of Aeronautics and Astronautics. In 1990, he joined the School of Electrical and Electronic Engineering at Nanyang Technological University, Singapore, where he is currently a tenured associate professor.

Dr Wang's current research interests include modelling and control of computer graphics rendering systems and also robust and reliable controls, nonlinear controls, and their applications to flight control systems. He has published 4 book chapters, about 70 journal papers, and more than 130 conference papers.

Dr Wang currently serves as an associate editor of *Transactions of the Institute of Measurement* and the *Asian Journal of Control*. He was a guest editor for a special issue of *Control and Intelligent Systems*. The special issue of this international journal published in January 2012 was dedicated to networked control and unmanned systems.

Dr Wang also served as the general chair of IEEE's 2007 International Conference on Control and Automation and program chair for the 2010 conference. He also chaired various aspects of several conferences including the International Conference on Control, Automation, Robotics, and Vision; the Asian Control Conference; the Chinese Control Conference; and others. He was named chairman of IEEE's Singapore Control Systems Chapter for 2008–2009. He is a senior member of IEEE.

1 Introduction

1.1 BACKGROUND AND MOTIVATION

While modern rendering software claims to have controlling mechanisms that enhance runtime performance, the mechanisms are often very primitive and inadequate. The results of this deficiency are indeterminate drops in the visual quality of generated imagery and frame rates that can severely affect usage experience. By applying control theories in real-time rendering, it is possible to rectify these shortcomings altogether.

The vision is to create an intelligent rendering system that can systematically adapt to its operating environment to produce optimum runtime performance at all times. To our best knowledge, no commercial product exists as this work is written and no active research is in progress in this cross-disciplinary application field.

The application of control concepts in the computer graphics software provides new opportunities for better performance derivable from graphics hardware. Until today, typical rendering applications struggled to utilise hardware efficiently. Much of the burden of optimisation falls on the software programmer who must be extremely conversant with the graphics pipeline.

The predominance of interactive computer graphics is underscored by a burgeoning variety of applications in various aspects of daily life. For example, it is easy to observe various types of interactive systems in an urban environment such as a shopping mall or an office building. These systems range from digital signage to projection-based displays and touch panels. At the industrial level, interactive computer graphics technology powers important processes such as computer-aided design and manufacturing, virtual prototyping, and scientific visualisation and simulation.

While customers constantly demand high quality computer-generated graphics, the cost associated with their demands may not be within reach. To illustrate, the price of a performance workstation is typically many times more than the cost of a desktop PC for home use. Furthermore, mobile devices such as PDAs and cell phones lack sufficient computing power to render high quality graphics for productivity at work.

Our research concerns a fully automated technology that circumvents the aforementioned problems and allows users to enjoy high quality interactive computer graphics on both desktop and mobile devices. The objective of this project is to leverage earlier research on this subject and extend the work to allow a product-ready toolkit to be developed for commercialisation opportunities.

Over the past few years, we developed a framework that realises the concept of delivering adaptive interactive rendering through laboratory experiments, theoretical modelling, and simulation. Our technology employs control theory and the system identification methodology, both of which are mature fields, proven by their use in aeronautical, mechanical and electrical engineering, and electronics industries. The concept is based on feedback control that can provide consistent performance monitoring and regulation with no requirement for human intervention. From a systems perspective, the technical challenge translates into the form of a "plant" (process to be controlled) and a "controller" component that ensures the process performs optimally according to predefined objectives.

This technology clearly has numerous applications and commercialisation possibilities. We conceived the possibilities listed below.

Computer-aided design (CAD) and -manufacturing (CAM)—Three-dimensional (3D) datasets used widely in many industrial applications. Our technology will allow a user to view such datasets even on a mobile device. This brings productivity out of the office and makes it available to people on the move.

Virtual communication—The market for 3D virtual communication is growing, particularly in the education and corporate services segments. As a viral social networking medium or mode of communication in professional exchanges, 3D interactive applications will remain key factors in online virtual communication. We see our technology as an enabling factor for linking more people to such networks.

Marketing and sales—More companies are moving toward high quality interactive content intended for consumers. This provides an opportunity for us to introduce our technology so that more people can utilise it without the limitations imposed by hardware. As a result, commercial entities can expect greater market reach and corresponding increases in revenue.

Training and education—Our technology can be deployed in various training and education products, enabling them to be delivered to audiences utilising hardware with different capabilities. The benefit offered by our technology is the easy ability to visualise 3D information even in a collaborative environment, therefore enhancing the value of training and knowledge dissemination.

1.2   OBJECTIVES AND CONTRIBUTIONS

Based on the shortcomings of current real-time rendering software, our research entailed the investigation and development of a feasible solution that would allow accurate and sustainable control of the real-time rendering process on different hardware platforms. The two key objectives affecting implementation of the technology are:

1. Despite the complexities involved in real-time rendering, it is imperative to devise a systematic method to describe this process in a form that relates its inputs and outputs consistently.
2. Based on the derivable form and the known characteristics of the rendering process, it is critical to find applicable control principles and frameworks that will ensure control of the process over a variety of scenarios.

Our research spans knowledge of the computer science (computer graphics rendering) and control engineering disciplines. Both fields imposed challenges that made our research both exciting and fulfilling. Our key research contributions are listed below.

1. We describe a novel framework by which the real-time rendering process may be modelled accurately. This framework involves the adoption of data-driven system identification methodology. Previous attempts to characterise the rendering process via only observable variables and case-specific formulations led to inaccurate models. Our model addresses these shortcomings.
2. Apart from linear models, our data-driven framework is extended to non-linear models using soft computing techniques such as neural networks and fuzzy models.
3. We developed control system frameworks for both linear and non-linear models in real-time rendering using (a) PID control with and without gain scheduling and (b) fuzzy control with and without adaptive neural networks.

The application of our control frameworks has shown much better resource utilisation in the real-time rendering process than earlier work that generally demonstrated coarse performance tracking.

1.3   SCOPE OF WORK

Real-time rendering is a vast topic in the field of computer graphics. Although the modelling techniques and control framework may be applicable to areas such as volume- and image-based rendering, our study deals with polygonal-based rendering pipelines found in commodity graphics hardware and it leverages geometry subdivision technique as a basis for controlling the input to the rendering system.

At this juncture, our work is based largely on the rendering of a single large 3D mesh that is used as a pseudo-representation of more complex 3D scenes with numerous objects. From a different perspective, this system is useful for applications involving a single large object of interest, for example, massive model rendering and computer-aided design.

Since the focus of this research is on real-time rendering relating to the response time of a system in an interactive environment, we consider the time required to render an image (frame) as the critical performance metric. While computer graphics activity is essentially visual, the quality of the generated image is frequently taken as the next most important metric for assessment. However, due to the subjectivity and complexity involved in processing image comparisons, the image quality component is omitted as a performance object in this work. From the system perspective, the real-time rendering framework proposed in this research is flexible to accommodate a multiple-input–multiple-output (MIMO) configuration. This means the user has the full freedom to implement additional output variables, which may include image quality related performance variables.

1.4   BOOK OUTLINE

Chapter 1 provides the background and motivation that led to this research. Chapter 2 discusses the fundamental knowledge in two key disciplines related to this research—real-time computer graphics rendering and system identification methodology. We then provide a systems perspective of the rendering process and explain the impacts of variables surrounding the system inputs and outputs. After that, a survey of previous research in the areas of rendering load control and characterisation is discussed.

Chapter 3 delves into the details of our data-driven modelling approach to real-time rendering with a focus on linear system structures and their derivation. Through experiments, we provide rendering models for single-input–single-output (SISO) systems and show how they may be extended to more complex and practical systems involving multiple inputs.

In Chapter 4, we explore the use of soft computing techniques for modelling the real-time rendering process. The application of such techniques is performed at the operating range of the rendering system where non-linear characteristics are exhibited. Following that, we provide the basis for linearisation from the derived non-linear rendering system model.

Chapter 5 begins with the introduction of model-based control and deals with the control system framework for the linear rendering system model obtained in Chapter 3. The key control mechanism discussed in this chapter is the closed-loop feedback design with PID controller. We demonstrate how systems with single and multiple inputs may be controlled as well.

The focus of Chapter 6 is on advanced control techniques and considers our proposed framework from a model-less perspective. This chapter illustrates the establishment of a control system framework without the need for an explicit system model as described in Chapters 3 and 4. By using a variety of fuzzy control techniques, we demonstrate that a control system can perform very well when tracking the performance of a real-time rendering process.

Chapter 7 discusses applications, challenges, and possibilities, including system architectures, software and hardware performance and future technology.

The conclusions of our research and suggestions for future work are discussed in Chapter 8.

Annex A contains sample applications.

Annex B discusses the authors' patent for Method and System for Adaptive Control of Real-Time Computer Graphics Rendering.

Annex C delineates Neural PID Control System Code.

2 Preliminaries

2.1 FUNDAMENTALS OF REAL-TIME 3D RENDERING

In real-time computer graphics, 3D rendering refers to the process of generating a sequence of images that produces not just the animated effect of motion and change but the visual cue of depth for objects in the imagery given an external input or stimulus to the system. In typical applications, the goal is to provide visual feedback to the user when there is interaction via the human-computer interface. The speed at which each image, known as a frame, of the animation sequence is generated defines the performance of the system.

Because speed of rendering every image is crucial in real-time rendering, both the computer hardware and software have to work together in the most optimal way so that the best possible image quality can be achieved in tandem with an acceptable frame rate (a metric that measures the number of frames that can be generated in one second). Over many years of research and development, the real-time 3D rendering process has taken leaps and bounds in terms of the image quality that is produced in various real-world applications such as computer games, training simulators and 3D product demonstrations. This involves an intricate process that spans the preparation of 3D content in elaborate modelling tools to processing combinations of rendering algorithms with myriad configurations of parameters for the final output which is the image to be shown eventually on the display device. Modern computers have dedicated hardware to handle computer graphics rendering. This hardware provides acceleration to computer graphics rendering routines so that the computer's central processor unit (CPU) can focus on other non-computer-graphics-related and auxiliary tasks. In general, real-time or interactive 3D rendering applications are supported by an abstraction layer that communicates with the hardware. This layer is commonly known as the 3D rendering Application Programming Interface (API) and it is fully responsible for pushing rendering commands to the hardware and managing the render state machine.

2.1.1 POLYGON-BASED RENDERING

Figure 2.1 shows the multi-stage 3D real-time rendering pipeline. The transformation of inputs to the final visible pixels on a display device may be described systematically via the following steps.

- **Creation in Local 3D Model Coordinate System**
  - Each object is created individually in its own 3D coordinate system.
  - Objects may be represented in a variety of geometry formats (triangles, rectangles, strips of polygons, etc.). Essentially, every polygon in a 3D space consists of points known as vertices.

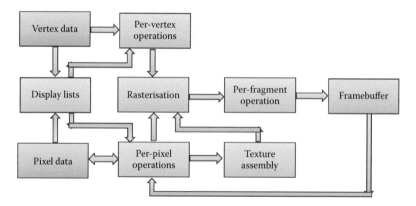

FIGURE 2.1   Real-time 3D rendering pipeline.

- For polygons to be rendered with visually correct features, each vertex is associated with a set of attributes such as position (coordinates in 3D space), colour, normal (perpendicular) vector from a surface, texture coordinates (user-defined mapping onto the surface), and other factors.

- **Transformation to Global World Coordinates**
  - To compose a scene in 3D space consisting of different objects, all created 3D objects must be transformed into the same coordinate system.
  - These transformations modify only the relative positions of the vertices and the normal. Visual attributes such as colour and texture coordinates are not modified.

- **Transformation to 3D View Coordinate System**
  - A viewpoint in 3D space is commonly cited as the "camera" location.
  - The geometry (vertex arrangement) from the 3D space is transformed into the camera view coordinate system. Depending on the rendering software, the common definition for this space is based on a right-handed coordinate system with the camera at the origin pointing down the negative z axis. The x axis is to the right and the y axis up. The projection from 3D to 2D space is performed at this stage.
  - The depth information of any object can be obtained from the z coordinate value at this stage.
  - The effect of virtual "lights" that create illumination properties in the 3D scene is computed at this stage. For example, a surface colour shading algorithm known as Gouraud shading will be computed at each vertex of a 3D object using the light parameters, light position, normal vectors, and the 3D object's texture or material properties.
  - The removal of polygonal surfaces not shown in the view due to occlusion is known as "culling" and is performed at this stage as well.
  - Culling is related to the attributes of the camera view defined by a virtual trapezoid volume known as the "view frustum" using six planes (left, right, up, down, front, and back) as shown in Figure 2.2.

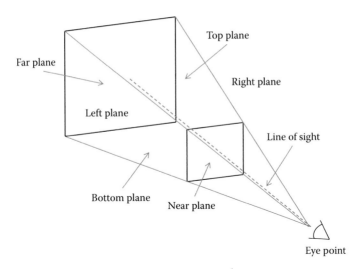

FIGURE 2.2 Camera view frustum in 3D space.

- **Transformation to 3D Clip Coordinate System**
  - The geometry data in this stage are prepared for a post-processing step known as "clipping."
  - The transformation of the geometry depends on the type of view projection used. Certain non-linear transformation may take place, for example, when perspective projection creates a tapering-off view of objects at a distant horizon in contrast to orthographic projection that consistently preserves the dimensions of a 3D object.

- **Transformation to Normalised Device Coordinates**
  - The geometry is normalised for display in a 2D window on a physical display device.
  - Further clipping is done to remove geometry outside the user-defined window boundaries.

- **Transformation to Display Window Coordinates**
  - All vertices are converted to units of the display (pixels) window.
  - Typically, the origin of reference is at the lower left corner of the display window.

- **Transformation to 2D Screen Coordinate System**
  - The conversion to screen pixels (rasterisation) is performed. Pixels are visible colour dots that can be displayed on a screen.
  - To generate shaded pixels, attributes such as texture coordinates, colour, and normal vectors are used in the computation and interpolated across the vertices and polygon surfaces.
  - Algorithms may be used to perform further hidden surface removal by using depth information obtained from the geometry.

- The final colour of the pixel is determined by combining all other effect state settings (e.g., blending and stencil operations) in the rendering pipeline.
- The output of this stage is the final colour of every pixel placed in the memory of the display hardware (the frame buffer).

In the course of rendering a 3D scene, many inputs and settings such as the geometries of 3D objects and their material "look" parameters are sent to the graphics hardware for processing. About a decade ago, outdated graphics hardware relied solely on a few hard-wired algorithms to process such data via a method known as the fixed function rendering pipeline. As a result, real-time rendering application developers had little space to control the look of a 3D object based on a limited set of functions that computed the rendering output. The impact of such limitations is the lower quality of imagery generated from computer graphics hardware.

This problem was circumvented by the advances represented by a new generation of computer graphics hardware that allows rendering routines known as *shaders* to be injected into the hardware before or during the runtime of an application. This capability now gives application developers full control over the quality of the generated output by varying shader routines. Figure 2.3 depicts this new-generation fully programmable rendering pipeline.

Shaders come in two formats: vertex and pixel types. A vertex shader is a graphics processing function used to add special effects to objects in a 3D environment. It is executed once for each vertex sent to the graphics processor. The purpose is to transform each vertex's 3D position in virtual space to the 2D coordinate at which it appears on the screen and the as a depth value in the graphics hardware. A pixel shader is a computation kernel function that computes colour and other attributes of each pixel. Pixel shader functions range from always outputting the same colour to applying a lighting value to adding visual effects such as bump mapping, shadows, specular highlights, and translucency properties. They can alter pixel depth or output more than one colour if multiple render targets are active. Figure 2.4 illustrates an example of the effects of pixel shaders on a 3D object. Apart from vertex and pixel shaders, an important feature of state-of-the-art graphics rendering architectures is the functionality of geometry shaders. Geometry shaders are added to the rendering pipeline to enable generation of graphics primitives, such as points, lines and different types of triangles after the execution of vertex shaders. With this capability, it is then possible to perform operations such as mesh resolution manipulation and procedural geometry generation.

Computer hardware technology and new rendering algorithms continue to advance quickly. The evolution of the real-time rendering pipeline also continues as this book is written.

2.1.2 VOLUMETRIC RENDERING

In Section 2.1.1, we described how animation can be produced using 3D data and physics-based principles for surface shading effects. Another technique for producing 3D visualization is through the usage of volume data that consists of not

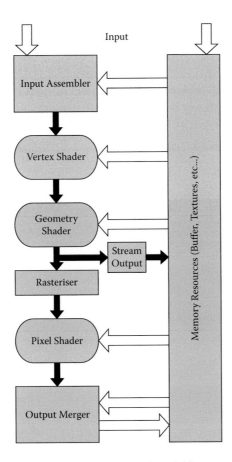

FIGURE 2.3 Programmable rendering pipeline (DirectX 11).

just positional information in 3D space but continuous depth data with additional dimensions and possibly its materials information as well. This type of spatial data is commonly used in scientific and medical work where cross-sectional information is important for evaluation and study. Volume rendering produces the exterior and the interior look of an object, usually with visual cues such as transparency and color differentiation. The image generation process considers the absorption of light along the ray path to the eye and volume rendering algorithms can be designed to avoid visual artifacts caused by aliasing and quantisation.

2.1.3 IMAGE-BASED RENDERING

In contrast to polygon-based rendering in which 3D geometry is provided for constructing the 3D hull of an object, image-based rendering techniques render novel 3D views by using a set of input images. This avoids the need for a stage where 3D data has to be explicitly provided by manual labour or some data acquisition means. These techniques focus on computer vision algorithms in feature detection and extraction from a set of basis images and thereafter reconstruct a 3D object or scene.

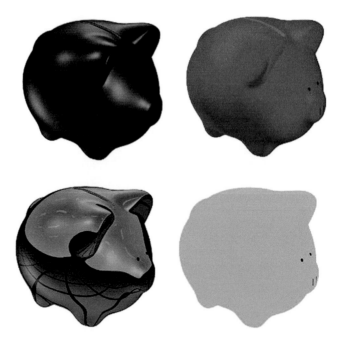

FIGURE 2.4 **(See colour insert.)** Samples of surface shading effects that can be achieved with pixel programs.

Image-based rendering techniques are often classified according to the degree by which geometry information is used. More importantly and in recent years, there has been a confluence of image-based techniques with polygon-based rendering in many applications due to the close continuum in 3D and 2D space in computer graphics.

As volume and image-based rendering are topics beyond the scope of this research, they are introduced here as auxiliary information on alternative 3D rendering techniques and more information can be found on the Internet and major research publication portals.

2.2   SYSTEM IDENTIFICATION

The goal of system identification is to derive a mathematical model of a dynamic system based on observed input and output data. Usually *a priori* information pertaining to a system will be useful for postulating the preliminary model structure. The system may then be modelled according to empirical data (black-box modelling) or conceivable mathematical functions such as physical laws (white-box modelling). Often, real world systems are non-linear and operate with reliance on state memory. The systems are dynamic and thus their outputs may depend on a combination of previous inputs, outputs, and states. The combination provides the basis for time series and regression mathematical expressions (models) for different reproducible systems.

System identification is an iterative procedure that can be summarised briefly by the flowchart in Figure 2.5. A model structure is chosen in advance based on

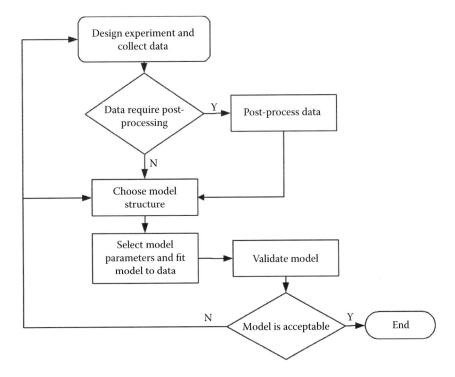

FIGURE 2.5   Process flow in system identification methodology.

preliminary information obtained from the system. The parameters of this model structure are then computed based on the set of experimental data collected previously. A portion of this data is allocated for model validation and the entire process from choosing a model structure to validation is repeated until the user-defined simulation performance criteria are met.

From a system identification perspective, we treat the real-time rendering process as the subject to be modelled. Since the rendering process cannot be described intuitively by physical laws such as mass, velocity, and temperature, black-box modelling is adopted. The system is first tested with a set of predefined inputs and the outputs are collected. This input–output dataset that captures a certain dynamic range of the behaviour of the system is then used with mathematical regression techniques to derive the estimated model.

Due to the scope of this book, we briefly summarise the steps in the system identification process below. A detailed and authoritative coverage of this topic can be found in Ljung's book [1].

2.2.1   DATA COLLECTION

To obtain an effective model of a system, it is necessary for the measured data to capture and show the behaviour of the system adequately. An appropriate experimental design can ensure that the correct variables and dynamics of the system are measured at sufficiently good resolution. In general, the following principles should be observed:

1. Select inputs that can excite the system dynamics adequately.
2. Minimise the effects of noise and disturbance to obtain a good signal-to-noise ratio.
3. Choose appropriate sampling intervals for measuring data.
4. Set a sufficient long duration of data collection to ensure capture of important time constants.

2.2.2 MODEL SELECTION

In system identification, we begin by determining the model structure best expressed by a mathematical relationship between input and output variables. This model structure typically provides the flexibility to describe a system based on certain parameters. Some examples of model structures include parameterised functions and state space equations. To illustrate, a linear parametric model is provided in the equation below.

$$y(k) = ay(k-1) + bu(k) \qquad (2.1)$$

where u is the input, y, the output, k, the discrete time step and a and b are model structure variables.

Essentially, system identification is a systematic approach that begins with the selection of a model structure and then using approximation techniques to estimate the numerical values of the model parameters. While it may seem arbitrary to start with the selection of a model structure, it is not an entirely ad hoc process. The following approaches may be adopted in deciding on an appropriate model structure.

1. Start with the simplest system model structures to avoid unnecessary complexity in cases where the data can be modelled by a simple structure. Alternatively, a user can try various mathematical structures in a technique known as black-box modelling.
2. Designate a specific model structure for the data to be modelled by establishing certain predetermined principles; this technique is known as grey-box modelling.

Some well known system model structures from established research include the:

Auto-regressive exogenous (ARX) model
Auto-regressive moving average (ARMA) model
Box–Jenkins model
Output error model
State space model

2.2.3 COMPUTING MODEL PARAMETERS

In system identification, the model parameters are estimated by minimising the function that describes errors between the derived system model output and the measured response. Assuming a system is linear and time-invariant, the output of the linear model y_{model} can be expressed as

$$y_{model}(t) = G(s)u(t) \tag{2.2}$$

where $G(s)$ is the transfer function, y the model output and u, the input to the model. To determine $G(s)$, we can minimise the difference between the model output $y_{model}(t)$ and the measured output $y_{meas}(t)$. We can use the minimisation criterion which is a weighted norm of the error $v(t)$:

$$v(t) = y_{meas}(t) - y_{model}(t) = y_{meas}(t) - G(s)u(t) \tag{2.3}$$

where $y_{model}(t)$ is either the model's simulated response given an input $u(t)$ or its predicted response given a finite series of past output measurements, i.e., $(y_{meas}(t-1), y_{meas}(t-1),\ldots)$.

From the above, $v(t)$ is otherwise known as the simulation error or prediction error. The objective of the estimation algorithm is to generate a set of parameters in the model structure G such that eventually this error is minimised.

2.2.4 EVALUATING QUALITY OF DERIVED MODEL

The steps taken to evaluate the quality of a derived system model generally include the comparison of the model response to the measured response and the analysis of model residuals. Figure 2.6 compares the outputs of two different models with a measured output.

Residuals are differences between a model's one-step-predicted output and the measured data. In other words, residuals may be understood as portions of validation data that are not well described by the model. In residual analysis, the whiteness and independence tests are key performance indicators.

The whiteness text examines whether a model includes a residual auto-correlation function inside the confidence interval of the estimates. If it does, the model passes the test and the outcome indicates that the residuals are not correlated.

In addition, a model is qualified when it passes the independence test (no correlation between its residuals and past inputs). If evidence indicates such a correlation, the information revealing how the output relates to the input is incomplete. A simple example is an output $y(t)$ beyond the confidence interval during a lag k that originates from the input $u(t-k)$. A good model should perform both tests relatively well.

The system identification methodology accommodates an iterative process in the determination of the final model structure and parameters. A real world system may not be represented by only a single model structure. Whenever a derived model is found inadequate, it is necessary to revisit the model selection process, reconsider

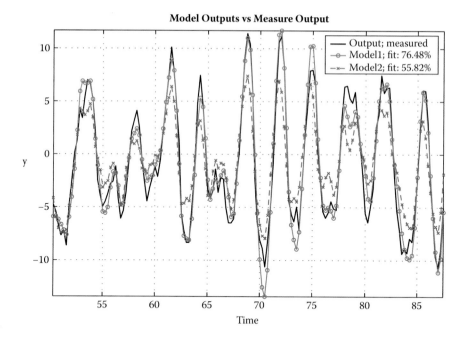

FIGURE 2.6 Comparison of two model outputs with measured system response.

the model parameter determination algorithm, and perhaps re-evaluate the data collection procedure.

2.3 LITERATURE REVIEW

While control theory is a mature field of study developed after the industrial revolution, the adoption of the techniques in the domain of computer software, particularly real-time computer graphics systems, remains nascent. This literature review provides a survey of research in these areas as background for our research.

2.3.1 COMPARATIVE STUDY ON EXISTING RESEARCH

The premise of the novelty in our research is founded upon close examination of previous work done in the fields of both real-time computer graphics and control theory, particularly those that have been successful in fusing the two disciplines and a careful thought process in terms of innovation in this area. A broad-stroke but systematic and progressive approach was taken to consider research publications within two decades to ensure that relevant techniques are not missed out regardless of their age and how they might contribute to further knowledge development.

Figure 2.7 shows the research comparative study flow conducted in this work which consists of the Classification Stage and the Qualitative Comparison Stage. In the Classification Stage, we begin with the most relevant keywords in the literature search terms. We consider the following words as the "lowest denomination"

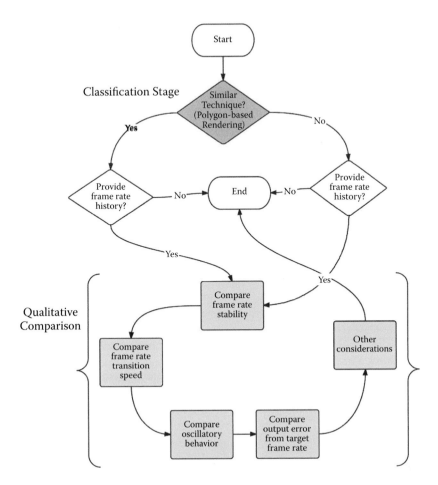

FIGURE 2.7 The comparative literature review workflow.

because of their relative importance in a subject matter. For example, omitting words such as "real-time", "graphics" and "3D" since they are either rhetorical in computer graphics research or they may be replaced by stronger keywords such as "interactive", "rendering" and directly meaningful candidates such as "frame rate" and "control". These keywords are used in search fields in major research publication online portals which indexes the world's largest collection of research literature. The gleaning process covers more than 500 research papers in a combined cohort of 4,000 search results from the publication portals.

As described in Section 1.3 in Chapter 1, the research in this thesis is primarily focused on polygon-based rendering technique which is predominant in common consumer and industrial applications such as computer games, virtual reality software and computer-aided design and prototyping. Hence, the Classification stage ends with segregating research literature that shares the same technique and is related to the topic of interactive 3D rendering. Table 2.1 shows the results from this classification stage from the initial pool of publications.

TABLE 2.1

Results from the Research Review Classification

	Polygon-based Rendering	Non-Polygon-based
With Frame Rate Data	[62] [64] [65] [67] [72] [74] [77] [81] [84] [85] [92]	[69] [73] [80] [83] [86] [87]
Without Frame Rate Data	[61] [63] [66] [68] [75] [78] [79] [82] [89] [90] [91]	[70] [71] [76] [88]

The next step in this comparative study is to select literature which provides experimental results on frame rate control since we need to conduct a qualitative analysis on them. For this purpose, these results should contain a history of data points in the time domain that demonstrate certain desirable qualities such as stability, offset errors and smooth frame rate transitions. These features would be compared to the results we obtain from the experiments conducted using the techniques proposed in this research with both qualitative and quantitative perspectives.

While research papers could be found relating to the topic of interactive rendering, however many of them alluded to concrete experiment results on sustainable performance as shown in the references from the bottom row of Table 2.1. In some cases [61] [71] [76], only static frame rates are given as an approximation to the interactive requirement. Furthermore, other researchers have chosen to work on volumetric [71] [76] [80] and image-based rendering [68] [70] [83] techniques which are prevalent in medical and large-scale visualization research but they differ from polygon-based rendering vastly. As a result, it is not straightforward to establish a direct comparison on the benefits offered by our research with these techniques. Despite these differences, we strive to provide a detailed qualitative and quantitative analysis on the aforementioned rendering architectures and their respective performance with our rendering framework in Chapter 8, Section 8.1.

2.3.2 Control-Theoretic Approaches to Computer Systems

As computer systems become increasingly complex through advances in hardware and software technology, traditional approaches to providing performance guarantees have become inefficient. In recent years, control engineering principles used successfully in real-world applications such as mechanical and electrical systems and process control have emerged as promising solutions to meet performance control challenges such as real-time scheduling, network bandwidth control, and power management in complex computer systems.

The comprehensive framework presented by Abdelzaher et al. [2] introduces feedback performance control in software services. The authors emphasised the importance of guaranteed quality of service (QoS) in modern computer software and systems that indicates the need for robust frameworks to achieve certain performance objectives. They further defined and explained the attributes of a QoS-aware service consisting of performance metrics such as queuing delays, execution latencies, and service response times. They also demonstrated that a software system can be approximated by a linearised model with corresponding conceptual software representations of actuators and sensors. Although the feedback control architecture was provided for generic software systems, the entire work focuses on web server applications.

Abdelwahed et al. [3] proposed a generic online control framework to design self-managing computer systems. The control actions governing system operations were obtained by optimising system behaviour as forecasted by a mathematical model over a specified time horizon. The case studies cited deal with power management under time-varying workloads and signal detection accuracy and latency levels.

Since computer systems in networked environments are gaining importance due to increasing Internet usage, Li and Nahrstedt [4] proposed a task control model to illustrate the dynamics of QoS adaptations using digital control theory. The objective was to provide optimum resource allocation to tasks in a distributed environment where multiple applications compete for and share limited system resources, thus ensuring the best user experience and efficiency. A proportional, integral, derivative (PID) controller was used to achieve the desired performance objectives relating to the QoS metrics.

Hellerstein et al. [5] provided a comprehensive overview of the challenges in control engineering of computer systems. Similar studies were reported by Abdelzaher et al. [7], Lu et al. [8], and Karamanolis et al. [9].

2.3.3 CONTROL PRINCIPLES IN COMPUTER GRAPHICS SOFTWARE

In Li and Shen's work [10], a fuzzy logic controller serves as an automatic mechanism for controlling error tolerance in hierarchical volume rendering. Volume rendering is a technique for directly displaying a sampled 3D scalar field without first fitting geometric primitives to the 3D discrete sampled date set. The performance criterion is a user-defined frame rate that the control system will strive to achieve based on adjusting the error tolerance.

Sort-last rendering is a computer graphics applications technique for rendering extremely large datasets in clusters of computers, usually in a distributed environment. Kirihata et al. [11] showed that it is possible to use feedback control to harness large data transfer processes in sort-last rendering.

Another example of the adoption of control principles in computer graphics software is the work by Dayal et al. [97]. They proposed an adaptive form of frameless rendering with the potential to increase rendering speed dramatically over conventional interactive rendering. This is done without the rigid sampling patterns of framed renderers and by allowing sampling and reconstruction to adapt with very fine granularity over spatial–temporal colour changes. A sampler uses closed-loop feedback to guide sampling toward edges or motion in the image to maximise rendering efficiency.

To date, little research has focused on the adoption of control principles in computer graphics applications related to rendering. While the potential benefits are immense based on a broader perspective in which control techniques have been used successfully in generic software, the challenges usually lie in specific applications that require in-depth understanding and appropriate modelling before the control concepts may be introduced.

3 Linear Model Analysis of Real-Time Rendering

3.1 INTRODUCTION

The real-time computer graphics rendering process embodies complex state transitions and fast dynamics amidst observable steady-state behaviour. To yield realistic or visually useful graphical information, the rendering process may be loaded with myriad combinations of the input variables and states to the point where it is important to describe this function in simple terms.

In this chapter, we describe the application of system identification methodology to real-time rendering. The basis for such an approach is that the rendering process may be treated from a system perspective as a data processing function. This allows us to analyse the process input and output data to establish some formal relationship between them.

3.2 BACKGROUND

The perennial and increasing demands for fidelity, coupled with hardware architecture advancements, pose many challenges to researchers and developers as they endeavour to find the optimal solution to accommodate both speed and quality of rendering. To this end, key techniques developed since the evolution of computer graphics three decades ago revolve around their ability to reduce the rendering load at application runtime. They are largely based on the principles of visibility reduction, geometry decimation, image-based methods, and more recently, techniques such as programmable shading.

As space does not permit an exhaustive review of these research efforts, we refer the reader to the comprehensive surveys by Cohen-Or et al. [12], Haines [13], and Akenine-Moller et al. [14]. Despite the ability of each approach to reduce rendering loads during runtime, their common weakness lies in the inability to guarantee stable frame rates.

In this chapter, we introduce a novel framework for obtaining an accurate model of an interactive rendering process. This framework is based upon the system identification methodology [1] that is a mature field of study associated with systems and control theory.

In addition to expanding understanding of the dynamics relating to the rendering process, the objective of modelling this process is to establish the groundwork for a control framework. Only with an accurate model can we design this control framework around the rendering process to yield the sustainable performance we desire.

In this research, we focus on exploiting a current trend in hardware technology that provides fine resolution in geometry control, known as *tessellation*. Since geometry is the primitive construct of any object in 3D space, it becomes a natural choice as one of the modelling variables in our framework. In brief, tessellation is the process of sub-dividing surfaces into smaller shapes with the objective of generating higher resolution information of the 3D model. Tessellation, also known as a subdivision technique, is a well researched field in computer graphics and had been adopted widely in many interactive rendering applications because of the visual acuity it provides. However, only recently has graphics hardware provided sufficient support for tessellation-based techniques in applications [30].

We introduce our modelling framework via experiments in two interactive rendering applications that use subdivision techniques in rendering load control. We aim to establish the fundamental validity of a system-based approach to modelling the rendering processes in applications similar to those selected in these experiments. Since rendering tasks are inherently complex in real-world applications, we provide a systematic extension from a single-input–single-output (SISO) model of the rendering process to a multiple-input–single-output (MISO) model that more closely resembles a broader class of applications. We hope that this progression along with the system modelling principles and fundamental considerations related to 3D rendering will enable readers to appreciate the value of this framework and acquire the necessary knowledge for its implementation.

Current research in rendering workload characterisation [16,17] and rendering time estimation [18,19] strives to profile the attributes of rendering without providing a systematic way to control the process. Often, the user is expected to arrive at some form of a primitive control strategy based on profile information. This often requires several attempts to re-evaluate control strategy and ad hoc refinement steps are often needed to remove major rendering bottlenecks.

This motivated us to attempt to utilise a systems perspective to model the rendering process. In this chapter, we demonstrate that accurate models can be obtained via our data-driven framework and extend this framework by introducing the use of a controller that can track and regulate frame rates with guaranteed performance. In comparison with other work, our research offers the following benefits:

- Our framework does not require the pre-processing of the 3D content utilised in other research [20,21,22]. Its performance is not limited to static pre-processed geometry and scenes.
- Our approach leverages hardware-accelerated technology (tessellation) that provides smooth transitions in geometry scaling unlike techniques that may generate visual hysteresis [21,22,23].
- The outputs of the derived rendering models exhibit very high accuracy when compared to actual rendering process outputs.
- When the derived rendering model is used in conjunction with a suitable controller, the closed-loop system can produce guaranteed frame rates. The self-correction process occurs entirely online during runtimes unlike current techniques that may require repetitive and labour-intensive offline evaluation.

3.2.1   CONTROL-CENTRIC DEFINITION FOR RENDERING TIME CONTROL

In contrast to previous research on interactive and time-critical rendering [20,22,24], we define rendering time control as a mechanism that should produce stable frame rates very close to the user-defined target instead of fluctuating below it. To date, much research on rendering time control has focused loosely on keeping the time required to render each frame within a certain budget and ignoring the quality of the control or the fluctuations resulting from the technique. This leads to two consequences.

The first implies that the times allocated to perform other tasks in an interactive application such as logic computation, collision detection, and animation will not be consistent. In some cases, "starvation" of other processes that require CPU or GPU resources may occur. This is detrimental to the effectiveness of visual simulation applications in which external devices that require CPU cycles are tightly coupled to the rendering process.

Second, weak frame rate control leads to suboptimal resource use. For example, an object rendered at 15 FPS that achieves acceptable visual quality should not be rendered at 25 FPS unless allowed by the user for valid reasons. This requirement is especially critical in interactive applications and systems with tight resource control policies such as in game consoles [25,26] and portable devices where sustainable and guaranteed performance is vital because processor time must be allocated for related non-graphics computations. In contrast, applying control engineering leads to analysis of system attributes such as output overshoot, settling time, and steady-state errors that constitute a better qualitative framework for performance monitoring. We feel that this is a more powerful outcome than the current research focus on frame rate control.

3.2.2   CHALLENGES IN USING HEURISTICS

Heuristics usually refers to an experience-based speculative formulation of a solution to a problem. Much research in the area of rendering performance control has been based on heuristics and analytical models [22,23,24,27]. As Gobbetti and Bouvier noted [24]:

> "...Static heuristics are not adaptive and are therefore inherently unable to produce uniform frame rates...."

Leukbe describes the difficulty in modelling the rendering process in his book on level of detail (LoD) for 3D graphics [28]:

> "...a predictive scheduler estimates the complexity of the frame about to be rendered... this approach is substantially more complicated to implement...because it requires an accurate way to model how long the specific hardware will take to render a given set of polygons."

The challenge in establishing reliable heuristics is straightforward. Driven by commercial demand and innovation, computer graphics hardware and software continue to change at unprecedented rates. In confirmation of this fact, Dumont et al. [29]

stated that given the complexity of real-rendering applications today, heuristics may fail in controlling rendering time. Haines [13] also describes this trend:

> "Perhaps one of the most exciting elements of computer graphics is that some of the ground rules change over time. Physical and perceptual laws are fixed, but what was once an optimal algorithm might fall by the wayside due to changes in hardware, both on the CPU and GPU. This gives the field vibrancy: we must constantly re-examine assumptions, exploit new capabilities, and discard old ways."

Based on these findings, dissecting the rendering process into distinct components that contribute to rendering cycle time is no trivial task. Tack et al. [18] did not consider overhead time in their performance model because of the complexity and additional costs it represented. The heuristics proposed in Wimmer and Wonker's work [19] varied in performance for different applications. This implies that unless an application is specially built to fit into their proposed framework it may not be easy to achieve stable frame rates across a broader range of applications.

Heuristics ignore non-linearity in their formulation, that is, they assume that functional relationships are always linear. This is unrealistic in practical applications because of the underlying hardware. Our experiments have shown that the time taken to render a vertex varies at different total processed vertex counts. The work of Lakhia et al. on interactive rendering [22] demonstrated that texture size has a non-linear relationship with the time taken to render a 3D object. Finally, heuristics face the same challenges as other frame rate control mechanisms in terms of balancing qualitative requirements such as visual hysteresis [23] and rendering performance.

3.2.3   PURPOSE OF WORKLOAD CHARACTERISATION AND ANALYSIS

Apart from heuristics in the quest to limit rendering time, researchers also analysed rendering workloads with the goal of identifying and eradicating bottlenecks at runtime. Kyöstilä et al. [16] created a debugger and system analyser for graphics applications running on mobile hardware. Monfort and Grossman [17] attempted to characterise the rendering workloads of 3D computer games via a specially developed tool. In recent years, major graphics hardware vendors have provided software toolkits that allow low level access to their hardware for debugging and in-depth analysis of graphics workload with the goal of optimising performance of interactive applications during runtime.

However, workload characterisation and analysis are not adaptive mechanisms that will bring about stable frame rates. They are helpful only for tracing bottlenecks and manifesting an application's rendering workload profile. To utilise these mechanisms for runtime performance, the process usually involves (1) identification of the problem (such as the cause of a bottleneck) during runtime followed by (2) manual effort to eradicate the bottleneck offline and then re-run the same scenario. This approach does not guarantee performance when the application use or 3D scene content changes. Since interactive rendering usually causes dynamic changes to visual content, the approach of using workload characterisation and tuning is not generally robust.

3.3   CASE FOR DATA-DRIVEN MODELLING

In system identification, we approach the problem of modelling a dynamic system from the observable data generated by its input and output. The case for using data-driven modelling is especially compelling for real-time rendering because the process is inherently complex. Rendering is a computer system process that thus raises considerations at both the hardware and software levels. Furthermore, unlike mechanical systems or chemical processes, no physical laws or intuitive functional relationships can be applied easily to achieve high accuracy.

Considering the real-time rendering process as a *black box* does not necessarily imply high risk of inappropriate modelling of the system as long as reasonable assumptions are based on a priori understanding of the system and can be reinforced from experiment results. In this book, we approach the challenge of modelling a rendering system by considering the expanded scopes of both single and multiple inputs. We also consider the output of the rendering process in terms of measurable quantities and the benefit of registering them as system outputs. This chapter discusses the inputs and outputs considered in system modelling and their eventual roles in system model representations.

To proceed with the modelling process, we first establish the relationship between the input and output of a system. This means that we must define and qualify the set of inputs and outputs before proceeding to identify their relationship. In the context of a real-time rendering application, it is reasonable to associate the geometry used for construction of 3D objects with the input to the rendering system and the output with the frame rate since empirical data indicate that they have an inverse relationship. Furthermore, in system identification, the input variables must be modifiable by the user in a straightforward manner. This is different from research in workload characterisation and heuristics where the defined variables are quantities such as hardware level parameters and processing time that cannot be changed by a user during runtime.

3.3.1   BASIS FOR SELECTION OF SYSTEM VARIABLES

With reference to the data flow in the computer graphics rendering pipeline shown in Figure 2.3 in Chapter 2, the inputs to the rendering process are obtained from memory resources (rectangle at far right) of the computer system. These inputs consist of various types of data ranging from geometry information to textures (image-related information) and rendering routines such as shader programs.

In order to define a set of variables to describe a rendering system, the input and output variables must be easily measurable. Furthermore, it is imperative that the input variables are controllable so that control actions can be implemented properly. Based on these criteria, we investigated the available performance counters with common low level graphics rendering profiler toolkits that included Microsoft's PIX. Table 3.1 shows a set of performance counters commonly used in many computer graphics applications.

Since many performance counters fall into the same category and are derivatives of one another, we chose the lowest denomination or most primitive variable in each

TABLE 3.1

Performance Counters in DirectX

Direct3D Counter Description	Official Name
FPS (#)	D3D FPS
Frame time in milliseconds	D3D frame time
Driver time in milliseconds	D3D time in driver
Triangle count (#)	D3D triangle count
Triangle count instanced (#)	D3D triangle count instanced
Batch count (#)	D3D batch count
Locked render targets count (#)	D3D locked render targets count
AGP/PCIE memory used in integer MB (#)	D3D agpmem MB
AGP/PCIE memory used in bytes (#)	D3D agpmem bytes
Video memory used in integer MB (#)	D3D vidmem MB
Video memory used in bytes (#)	D3D vidmem bytes
Total video memory available in bytes (#)	D3D vidmem total bytes
Total video memory available in integer MB (#)	D3D vidmem total MB

Source: NVPerfKit documentation from www.nvidia.com

selected category. To illustrate, the input geometry to the rendering pipeline may include lines, triangle fans, strips, and polygons. These are different input formats that share the same basis—3D geometry data. Hence the natural choice as the input variable of a rendering system should be the vertex count.

In addition to finding the appropriate variable by using its simplified form, another very important characteristic that determines suitability is whether a variable can be changed easily. For example, the batch counts and batch sizes of indexed buffers can impact the performance of a rendering system. However, little can be done to control these variables during an application runtime because these batches of vertices are predefined.

Finally, the resolution at which the selected variable may be adjusted affects the quality of the system model as well. The ideal case would involve a variable that allows fine resolution changes. For example, since the number of vertices is used as an input variable of a rendering system, it may be difficult to obtain an accurate model when this number can be varied only in limited steps.

One reason for this limitation is the underlying geometry LoD mechanism that controls the resolution of a 3D object with a certain topological objective and algorithm. The discrete LoD technique is an example of such a mechanism. Figure 3.1 illustrates the progressive variation (in steps) in the number of vertices that describe a 3D object. Conversely, other techniques such as progressive meshes and geometry tessellation allow 3D geometry variation at fine resolution levels. These techniques are preferred in comparison to the approaches cited earlier.

So far we have discussed guidelines for inputs to the rendering system. As for the output of the rendering system, the performance metric of primary concern to a user of real-time computer graphics is widely accepted as the frame rate (inverse of the time required to render one frame or image in a sequence) and quality of the generated imagery. The frame rate has a significant impact on the quality of the visual

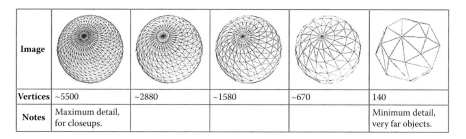

Image					
Vertices	~5500	~2880	~1580	~670	140
Notes	Maximum detail, for closeups.				Minimum detail, very far objects.

FIGURE 3.1   Visual effect of varying vertex count for 3D object in discrete steps. (Source: http://en.wikipedia.org/wiki/Level_of_detail#A_discrete_LOD_example)

experience offered by a real-time rendering application. While the quality of the generated imagery may be important to the user, the interactive experience is usually dominated by the application response rather than the quality of the generated imagery. Furthermore, quality is a subjective notion that complicates the adequacy of any useful metric.

3.4   LINEAR SYSTEM MODEL REPRESENTATION FOR REAL-TIME RENDERING

This section describes the modelling process applied to the real-time rendering system and the derivation of the mathematical models for various types of rendering applications. Using the system identification methodology, we demonstrate that linear time-invariant models can be obtained from the input and output data collected from experiments conducted using sample rendering applications.

A basic relationship between the input and output of a system may be expressed as a linear difference equation as follows.

$$y(t) + a_1 y(t-1) + \ldots a_{n_a} y(t-n_a) = b_1 u(t-n_k) + \ldots + b_{n_b} u(t-n_k-n_b+1) + e(t) \quad (3.1)$$

where:

$a_1 \ldots a_{n_a}$ and $b_1 \ldots b_{n_b}$ are parameters to be estimated.
$y(t)$ is the output of the system at time t.
$y(t-1)$ and $y(t-n_a)$ are the previous outputs on which the current output depends.
$u(t-n_k)$ and $u(t-n_k-n_b+1)$ are the previous inputs on which the current output depends.
n_a is the number of poles of the system or the order of the system.
n_b represents the number of zeroes plus one.
n_k denotes delay in the system.
$e(t)$ equals noise.

An alternative way to represent Equation (3.1) in a more compact manner is the ARX model described below:

$$A(q) y(t) = B(q) u(t-n_k) + e(t) \quad (3.2)$$

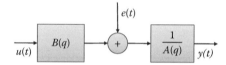

FIGURE 3.2   ARX model structure.

where q is the delay operator and $A(q)$ and $B(q)$ are represented as:

$$A(q) = 1 + a_1 q^{-1} + \cdots + a_{n_s} q^{-n_s} \tag{3.3}$$

$$B(q) = b_1 + b_2 q^{-1} + \ldots + b_{n_b} q^{-n_b+1} \tag{3.4}$$

and q is known as the backward shift operator defined by $q^{-1}u(t) = u(t-1)$. Figure 3.2 shows a graphical representation of an ARX model. In the context of this research, $u(t)$ and $y(t)$ may be taken as the input vertex count and output frame rate, respectively. For compact notation, the following vectors are used:

$$\theta = \begin{bmatrix} a_1 \ldots a_{n_a} & b_1 \ldots b_{n_b} \end{bmatrix}^T \tag{3.5}$$

$$\varphi(t) = \begin{bmatrix} -y(t-1) - y(t-2) \ldots - y(t-n_a) \\ u(t-n_k) \ldots u(t-n_k-n_b+1) \end{bmatrix}^T \tag{3.6}$$

From Equations (3.5) and (3.6), Equation (3.2) can be expressed as:

$$y(t) = \varphi^T(t)\theta \tag{3.7}$$

Alternatively, we can use the following notation to highlight the dependency of $y(t)$ on the set of parameters in θ:

$$\hat{y}(t \mid \theta) = \varphi^T(t)\theta \tag{3.8}$$

We want to compute the set of parameters θ by using the least square method and the criterion function:

$$V_N(\theta, Z^N) = \frac{1}{N}\sum_{t=1}^{N}\frac{1}{2}[y(t) - \hat{y}(t \mid \theta)]^2 = \frac{1}{N}\sum_{t=1}^{N}\frac{1}{2}[y(t) - \varphi^T(t)\theta]^2 \tag{3.9}$$

with the objective to get:

$$\hat{\theta}_N^{LS} = \arg\min_{\theta} V_N(\theta, Z^N) = \left[\sum_{t=1}^{N}\varphi(t)\varphi^T(t)\right]^{-1}\sum_{t=1}^{N}\varphi(t)y(t) \tag{3.10}$$

where $Z^N = [y(1), u(1), y(2), u(2) ..., y(N)$, and $u(N)]$ are the set of recorded inputs and outputs over a time interval of $1 \leq t \leq N$. Since the parameters of the model are encapsulated in the vector θ, solving Equation (3.10) gives us their numerical values.

Alternatively, a model may be represented in the state–space format whereby the inputs, outputs, and state variables are expressed as vectors and the differential and algebraic equations are written in matrix form. This format provides a convenient and compact way to model and analyse systems with multiple inputs and outputs. The state–space representation of a discrete time-invariant dynamic system model is described by the equations below.

$$x(k+1) = Ax(k) + Bu(k) + Ke(k) \qquad (3.11)$$

$$y(k) = Cx(k) + Du(k) \qquad (3.12)$$

where $x(k)$ is the state vector, $y(k)$ is the system output, $u(k)$ the system input, and $e(k)$ the stochastic error. A, B, C, D, and K are the system matrices. The derivation of the system matrices can be found in common system modelling and control engineering textbooks such as [1].

Although many model structures are used in the system identification field we consider primarily the two structures described in this section because of the advantages they offer in comparison to other model structures. The ARX model structure offers computational efficiency in polynomial estimation. It is thus preferable in many situations, particularly when model order is high. On the other hand, state–space equations provide mathematical constructs that leverage linear, first-order derivative variables that allow convenient computation even for systems involving multiple-input–multiple-output systems.

3.5   EXPERIMENTS

This section discusses two experiments conducted to derive the system model of rendering processes and one experiment illustrating the use of a derived model in a control system.

3.5.1   EXPERIMENT 1: SINGLE-INPUT–SINGLE-OUTPUT (SISO) SYSTEM

To illustrate how our modelling framework may be applied, we select two applications that make use of geometry subdivision techniques for rendering 3D objects with high visual details. These two applications are taken from a popular 3D graphics rendering toolkit (NVIDIA DirectX SDK) designed to help software developers exploit this subdivision technique in current computer hardware. The choice of the two applications is based on the current trend [30] for subdivision techniques in many application domains. The objective was to show that our framework can generate system models that describe the selected rendering applications accurately.

In Experiment 1, we wanted to establish a SISO rendering system model based on a sample interactive rendering application. The input and output variables were chosen as the input geometry (number of vertices) and frame rate, respectively. We selected the N-patch tessellation application (Figure 3.3) from DirectX SDK that

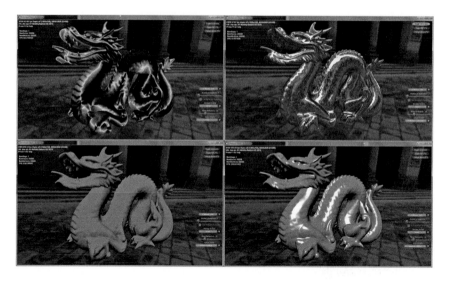

FIGURE 3.3 **(See colour insert.)** Screenshot of hardware tessellation sample application from DirectX SDK adapted with Stanford Dragon model in Experiments 1 and 2.

allows a user to set the tessellation segment values interactively to see the effect on the 3D object and the frame rate. To simulate the rendering quality in practical application, we modified the system for this experiment to load an environment map and apply different shading effects on 3D objects. The application was developed to allow easy configuration—a feature that makes it applicable for massive model rendering.

We collected two sets of data including the vertex input values and frame rates over six tessellation segments. The first set of data consisted of inputs between 120,000 and 920,000 vertices and frame rates (outputs) between 122 and 24 FPS over approximately 3,500 consecutive rendered frames. The second set consisted of 7,000 rendered frames with the same input and output variables. Approximately 5,200 frames were used to derive the model; the remaining frames were used for validation.

In addition to deploying the modelling framework on this application, we extended the validation of the framework to a computer game with highly complex 3D rendering. Figure 3.4 is a screenshot of the "Crysis" computer game (© Crytek). A total of 22,000 frames of input and output data were collected using Microsoft's PIX performance profiling toolkit. The first 20,000 frames were used to derive the model and the remaining frames served as validation data.

3.5.2 EXPERIMENT 2: MULTIPLE-INPUT–SINGLE-OUTPUT (MISO) SYSTEM

Since interactive rendering applications involve complicated processes, we extended our modelling framework to support multiple inputs in Experiment 2. In contrast to previous research [20,21,24] that did not focus on shading (an important aspect of 3D rendering), we demonstrated through Experiment 2 that more than one property may be captured by our framework. In particular, we developed a shader program that accepts a numeric value between 1 and 6 to control the quality of the surface

FIGURE 3.4   (See colour insert.) Screenshot of application in Experiment 1.

shading of the 3D object in the application (Figure 3.3). From a causal system perspective, this variable is selected because of its independence from the geometry input variable. No tight coupling exists between the shader complexity variable used in the raster stage of the rendering pipeline and the input geometry typically used in the pre-rasterisation stage.

In the data collection process in Experiment 2, we manipulated the vertex input to the rendering process by changing the tessellation segments and the shader complexity values at various tessellation levels. The corresponding frame rate changes were registered. Of the 14,000 frames collected, 12,000 frames were used for model derivation and the remainder for validation. From Experiment 2, the derived model allowed us to extrapolate the framework to support multiple inputs in more complicated rendering processes.

3.5.3   Experiment 3: Control Framework Using System Model

The objective was to construct a simple control framework using a system model derived from the aforementioned modelling process. We selected another application from the DirectX toolkit that used a progressive mesh control mechanism (Figure 3.5). The technique is similar to geometry tessellation and has been adopted widely in many interactive graphics applications to achieve fine resolution control of a 3D object's geometry. After a model of the rendering process was derived, we introduced the concept of a controller to manage the input to the rendering process to produce a rendering framework that offered stability and conformed to a user-defined frame rate.

In Experiment 3, we collected 60,000 frames of data; 50,000 were used for model derivation and the remainder for validation of the system model. The input data

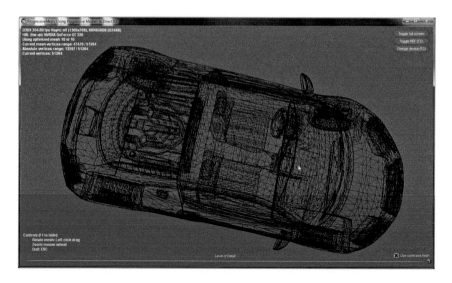

FIGURE 3.5　**(See colour insert.)** Screenshot of application in Experiment 3.

ranged from 38,000 to 45,000 vertices and the frame rate (output) between 330 and 400 FPS (see Figure 3.13).

After we obtained the rendering system model in MATLAB®, we exported it to Simulink® for control system design. In Simulink, we constructed a feedback loop and introduced a PID controller to correct the error between the simulated model output and the user-defined performance target (frame rate). In a similar way, this architecture may be implemented in interactive rendering software to achieve a constant frame rate.

All experiments were run on a desktop computer with an Intel Core2 Quad CPU at 3 GHz, with 8 GB of main memory and NVIDIA GeForce GT 320 graphics processor hardware (with 4 GB video memory) on a 64-bit Windows 7 operating system. Since the experiments were run on a generic Windows PC, we were aware of the system processes that may have shared the computing resources during data collection. To best preserve the integrity of the experiment data, changes of inputs to the rendering process were introduced programmatically rather than via mouse input. The system identification toolbox [31] in MATLAB/Simulink was used for modelling rendering processes in all the experiments.

3.6　RESULTS

3.6.1　Experiment 1

Using the sample application from DirectX SDK as shown in Figure 3.3, we can observe from Figure 3.6 that as the input vertex count increases due to greater tessellation, the frame rate decreases. This relationship is further plotted in Figure 3.7 where the impact on frame rate due to vertex count increase is shown.

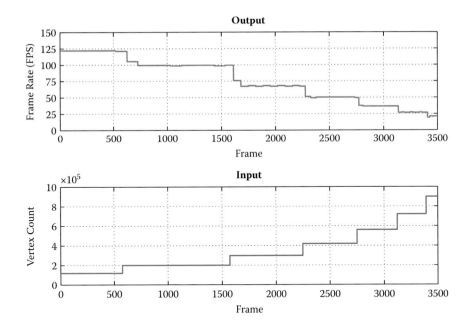

FIGURE 3.6   Input and output profiles of application in Experiment 1.

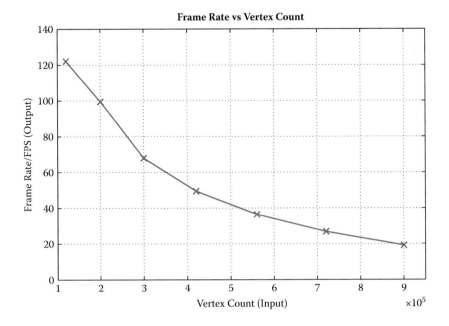

FIGURE 3.7   Steady-state frame time and vertex count relationship in Experiment 1.

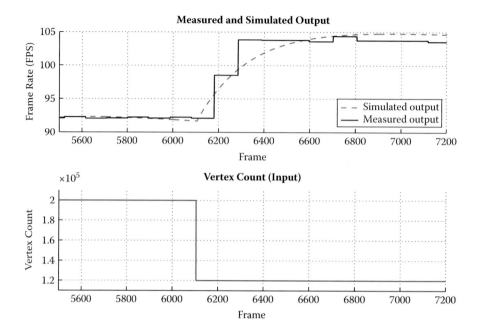

FIGURE 3.8 Measured and simulated output of rendering application in Experiment 1.

For an ideal linear system, the hardware's processing capability should remain unchanged at all operation ranges. This means that the time taken to process every vertex should be constant. However, we can see in Figure 3.7 a non-linear trend in vertex processing time as vertex count increases. The relationship may be approximated with multi-linear segments that fit the curve shown in the same figure. With this prior knowledge of the model, we proceed with model identification based on data measured in Experiment 1.

The simulated output of the derived model is shown in Figure 3.8. Note that the derived model produced reasonably accurate results in comparison to measured data with a maximum error less than 5 FPS in steady-state and best fit value of 84%. The best fit computation is:

$$Best\ Fit = \left(1 - \frac{|y - \hat{y}|}{|y - \bar{y}|}\right) \tag{3.13}$$

where y is the measured output, $\hat{y}$ is the simulated output, and $\bar{y}$ is the mean of y. A 100% value corresponds to a perfect fit. This result validates our hypothesis that the range is approximately linear. The model parameters are estimated in MATLAB using the system identification toolbox; the values are provided in Table 3.2.

In the second part of Experiment 1, we modelled the rendering of computer game software as shown in Figure 3.4. Figure 3.9 shows the measured and simulated outputs of the system. It is noteworthy that the environment to be modelled becomes

TABLE 3.2

Parameters of ARX Model in Experiment 1

Parameter	Calculation
A(q)	$1 - 2.444\,q^{-1} + 2.001\,q^{-2} - 0.5565\,q^{-3} + 4.21e - 006\,q^{-4} + 0.1708\,q^{-5}$ $- 0.4541\,q^{-6} + 0.2823\,q^{-7}$
B(q)	$-2.375 \times 10^{-7}\,q^{-1} + 3.072 \times 10^{-7}\,q^{-2} - 2.84 \times 10^{-7}\,q^{-3} - 3.156 \times 10^{-7}\,q^{-4}$ $- 1.18 \times 10^{-7}\,q^{-5} + 1.263 \times 10^{-6}\,q^{-6} - 6.156 \times 10^{-7}\,q^{-7}$
Operating point	$u = 1.5331 \times 10^5,\ y = 98.261$

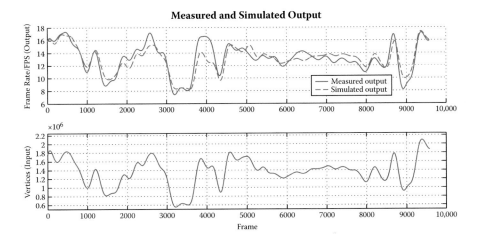

FIGURE 3.9   Error between measured and simulated output of application in Experiment 1.

more difficult as shown in the varying output levels compared to the first experiment. The actual data captured consist of more spikes due to interruptions from other computer processes and the data must be filtered so that a reasonable model can be derived. Nevertheless, through the proposed approach, we are able to obtain a system model that produced output with an error less than 4 FPS. The parameters of this model are presented in Table 3.3.

3.6.2   EXPERIMENT 2

We extended our modelling framework for rendering to consider more than one input. Based on selected combinations of two input variables (vertex count and shader value), we generated steady-state output responses of three settings as shown in Figure 3.10. Each graph in the figure indicates the steady-state input–output relationship exhibited by the system based on a certain combination of the values of the two inputs. The profiles of the measured inputs and outputs of the actual rendering are shown in Figure 3.11. A comparison of the simulated model and the measured

TABLE 3.3

Parameters of State Space Model in Experiment 1

		x1	x2	x3	x4	x5
A	x1	0.99984	−0.0057622	0.0037499	−0.00077957	0.0038575
	x2	0.0055967	0.99989	−0.011718	0.0050316	0.0021258
	x3	−0.0037656	0.011556	0.99982	0.010514	−0.0002729
	x4	0.00092734	−0.0051813	−0.010446	0.99965	−0.024729
	x5	−0.0035811	−0.0020073	0.00024635	0.024629	0.98977
	x6	−0.00046382	0.0023359	−0.0002931	−0.0015093	−0.040291
	x7	0.00070039	8.827e−005	0.00030079	−0.00013185	−0.0090756
	x8	0.0010839	−0.0010905	0.0004099	0.00070348	−0.0029228
	x9	0.0027148	0.0029469	0.0012923	−0.00086137	−0.007816

		x6	x7	x8	x9
	x1	−0.00011395	−0.0017801	0.00070025	0.0010708
	x2	0.00016522	−0.00052253	1.6105e−005	−2.5063e−006
	x3	−1.3826e−005	1.2645e−005	1.5497e−006	−3.1376e−007
	x4	−0.0020448	0.0045989	0.00012686	−7.5176e−006
	x5	0.068805	0.26254	−0.033504	−0.022596
	x6	1.0012	−1.3995	0.28977	−0.179
	x7	0.091242	0.48047	−0.049181	−0.16667
	x8	−0.28184	0.43031	0.18121	−0.94739
	x9	0.63828	−2.6907	0.44516	0.19954

		u1
B	x1	−5.4522e−007
	x2	9.7623e−008
	x3	6.6177e−009
	x4	−1.0667e−006
	x5	−1.0271e−005
	x6	2.1522e−005
	x7	0.00016518
	x8	0.00089275
	x9	0.00023178

		x1	x2	x3	x4	x5
C	y1	−138.38	5.7191	−3.9296	−20.813	−0.72794

		x6	x7	x8	x9
	y1	0.054266	−0.044465	−0.097836	−0.05148

		u1
D	y1	0

		u1
K	x1	3.2518
	x2	0.84349
	x3	−3.5836

TABLE 3.3 ***(Continued)***
Parameters of State Space Model in Experiment 1

		u1
	x4	–21.457
	x5	18.773
	x6	−1.8395
	x7	–0.85923
	x8	–0.89561
	x9	1.1259
Operating point	u = 1.4552 × 10⁶, y = 14.6932	

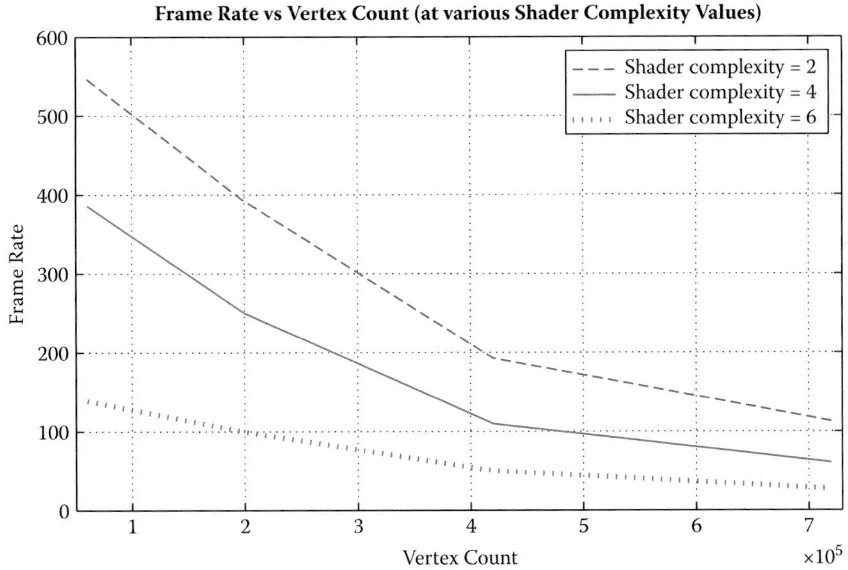

Frame Rate vs Vertex Count (at various Shader Complexity Values)
Shader complexity = 2
Shader complexity = 4
Shader complexity = 6
Frame Rate
600
500
400
300
200
100
0
1
2
3
4
5
6
7
Vertex Count
×10^5

FIGURE 3.10 **(See colour insert.)** Steady-state outputs of the system based on selected combinations of two input variables.

output is shown in Figure 3.12. We can observe that the simulated output tracks the measured output very closely—generally less than 2 FPS. Table 3.4 illustrates model parameters.

In Figure 3.11, the top and middle diagrams show the variations of the inputs to the rendering system while the bottom diagram shows the corresponding changes in the output. It can be seen that both inputs are varying during the experiment and none is held constant. This is to ensure that the data captured is representative of a MISO system model. In Figure 3.12, the top diagram shows the comparison of the output of the system model derived from the experiment data and the actual measured output. It can be observed that the simulated model output is very

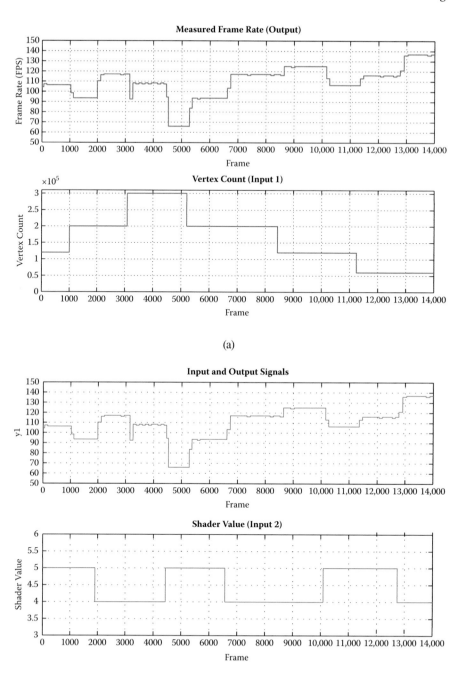

FIGURE 3.11 Profiles of two inputs and output of rendering system in Experiment 2.

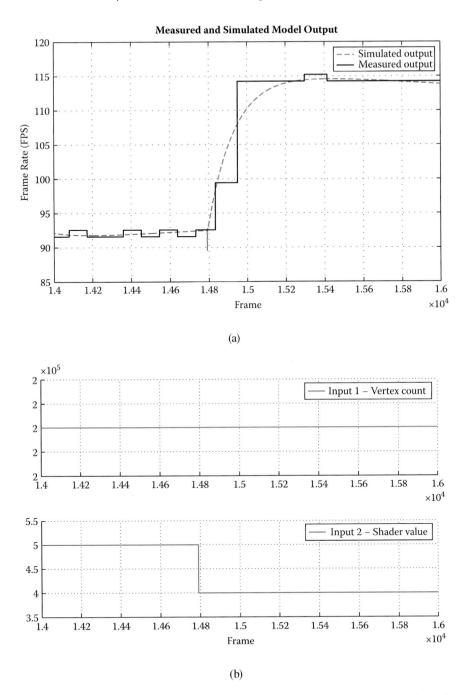

FIGURE 3.12   Measured and simulated outputs of MISO rendering system in Experiment 2.

TABLE 3.4
Parameters of ARX Model in Experiment 2

Parameter	Calculation
A(q)	$1 - 1.937\,q^{-1} + 1.019\,q^{-2} - 0.2182\,q^{-3} + 0.1363\,q^{-4}$
B1(q)	$-4.702 \times 10^{-5}\,q^{-1} + 4.257 \times 10^{-5}\,q^{-2} + 5.496 \times 10^{-5}\,q^{-3} - 5.051 \times 10^{-5}\,q^{-4}$
B2(q)	$2.918\,q^{-1} - 8.402\,q^{-2} + 7.855\,q^{-3} - 2.37\,q^{-4}$
Operating point	$u1 = 1.7839 \times 10^{5}$, $u2 = 4.3704$, $y = 109.0672$

close to the actual measured data signalling a highly accurate system model. The diagrams in the middle and bottom are snapshots of the inputs corresponding to this measured output.

3.6.3 EXPERIMENT 3

The objective was to adapt our modelling framework to another application and more importantly demonstrate the possibility of constructing a control system that provides stable frame rates based on this system model. Again, we first derived the rendering process model using experiment data collected from the application. Figure 3.13 illustrates a profile of this data. After a suitable system model

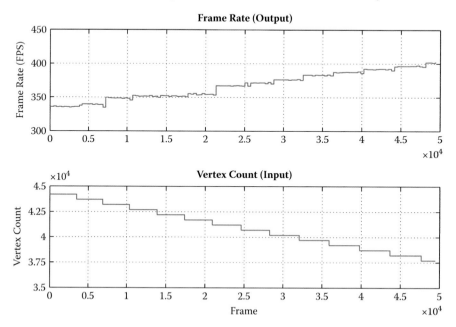

FIGURE 3.13 Profiles of input and output of rendering system in Experiment 3.

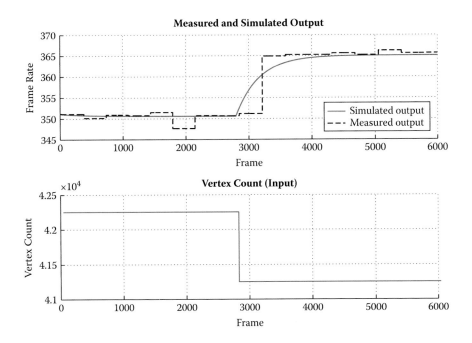

FIGURE 3.14   Measured and simulated rendering system output in Experiment 3.

TABLE 3.5
Parameters of ARX Model in Experiment 3

Parameter	Calculation
A(q)	$1 - 4.159\,q^{-1} + 5.814\,q^{-2} - 1.659\,q^{-3} - 3.322\,q^{-4} + 3.158\,q^{-5} - 0.8321\,q^{-6}$
B(q)	$1.007 \times 10^{-5}\,q^{-1} - 5.033 \times 10^{-5}\,q^{-2} + 0.0001007\,q^{-3} - 0.0001008\,q^{-4}$ $+ 5.045 \times 10^{-5}\,q^{-5} - 1.011 \times 10^{-5}\,q^{-6}$
Operating point	$u = 4.1809 \times 10^{4}, y = 356.7121$

was derived, we compared the output of the model with the actual measured data as shown in Figure 3.14. Similar to the comparison of the outputs in Figure 3.12, it can be observed that the simulated model produced an error rate less than 5 FPS throughout the simulated range. Table 3.5 illustrates model parameters.

Subsequently, we imported this model into Simulink and constructed a PID-based controller system as shown in Figure 3.15. We present the performance of this control system based on its tracking a pre-defined output level as shown in Figure 3.16. Note that the rendering system output follows the user-defined reference very closely at steady state and within a very short time without overshoot or oscillation.

To further validate our control framework, we replaced the system model with the actual rendering process in a separate test. The PID controller block was executed

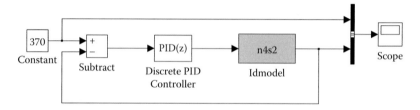

FIGURE 3.15 **(See colour insert.)** SISO control system in Experiment 3.

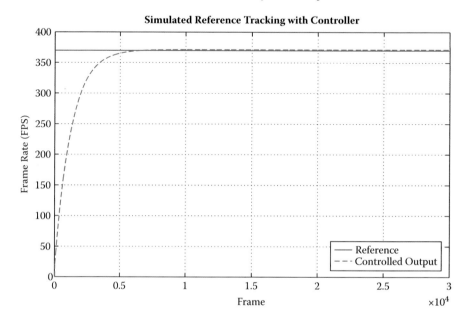

FIGURE 3.16 Simulated reference tracking with PID controller.

on a different computer and communicated with the rendering process via a network link using the transmission control protocol (TCP). The purpose of such a modified design was to avoid the interference and loading of the rendering process from MATLAB computation. The output of this experiment is shown in Figure 3.17.

Note from Figure 3.17 that the actual rendering application tracks the pre-defined reference level accurately as in the previous case. The steady-state error is negligible (less than 1% of the reference value), further reinforcing the validity of our system model and control framework.

3.7 DISCUSSION

One challenge we faced in this research was the stability of frame rates during data collection. We noted that frame rates on certain computers fluctuated even without changes in input geometry or user-controlled events such as mouse movement

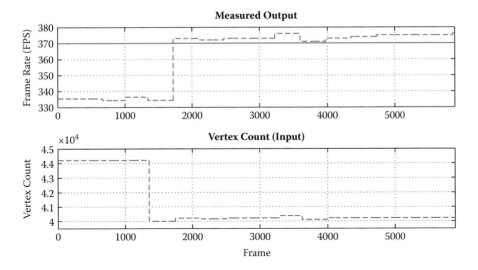

FIGURE 3.17   Reference tracking with actual rendering application.

and keyboard input. Furthermore, in our analysis of the dynamics of the rendering process, we observed various step-like interruptions in the frame rate changes that persisted for a certain period. We attribute this phenomenon to hardware or software operations such as driver wait states and memory transport delays. Furthermore, current tools available to us do not allow access to sufficiently low-level operations to identify these interruptions and irregularities.

As indicated in previous research [32,33,34], common performance metrics in interactive rendering affect mainly frame rate and image quality. In this research, frame rate was selected over image quality because it is well known that computation at image level is resource intensive [29] and assessment of image quality metrics may be subjective. Furthermore, research indicates that frame rates exert greater impact on user perception and response [35,36,37] in many applications.

As noted in previous research [24,27] that reactive rendering time control techniques cannot deal well with abrupt changes in scene load because the resultant oscillations in frame rates may negatively affect the user experience. We have shown in this research that our data-driven modelling framework provides an important basis for application of a control strategy that generates stable frame rates without noticeable oscillation. These are the benefits of employing system modelling and control techniques in the real-time computer graphics domain.

3.7.1 Comparison with Other Estimation Techniques

Many application developers may intuitively perceive the real-time rendering process as linear and thus use simplistic mathematical models to estimate its performance. In this section, we validate the accuracy of our data-driven modelling framework by comparing its outputs with those of two other intuitive estimation models based on a general relationship between the input and output of the rendering system. They are formulated mathematically as shown in Equation (3.14):

$$t_{frame} = k \times m_{vc} \tag{3.14}$$

where t_{frame} is the time taken to render a frame and m_{vc} is the rendering load represented by the total number of vertices used in rendering the 3D scene. In the first formulation, we want to express k as a factor estimated from all the training data. In the second formulation, k is expressed as an average of the previous n rendered frames. Based on this formulation, we may represent Equation (3.14) as a single best-fitting line segment expressed as:

$$f(x) = p_1 x + p_2 \tag{3.15}$$

where $f(x)$ is the function describing the line segment, p_1 is the gradient of the line and p_2 is the vertical axis intercept. Hence k in (3.14) and p_1 from above may be associated directly as:

$$k = p_1 \tag{3.16}$$

Using the curve fitting technique from the MATLAB toolbox, we obtain the line segment for the operating range in Experiment 1 with p_1 as 5.166×10^{-8}. With reference to the experiment data in Figure 3.7, given the input vertex count of 120,000, the estimated frame time using the value of p_1 is 0.0061992 s. This translates to a frame rate of 161.3111. However, the measured frame rate is approximately 104, yielding the error from this formulation as 57 FPS—in stark contrast to our model's output that is much more accurate. Over the entire tested range, the error between our model's output and the measured output is less than 5 FPS.

Next, we want to compare our system model with another that takes into account the k factor for previous frames instead of a single k factor for estimating frame time at any input point. Mathematically, this second model can be expressed as the n-moving average s given a sequence $\{a_i\}_{i=1}^{N}$ taking the average of n terms.

$$s = \frac{1}{n} \sum_{j=i}^{i+n-1} a_j \tag{3.17}$$

Therefore, each term a in the context of Equation (3.17) is the k factor estimated from a window of x number of frames. This gives us a set of values for k over the test range. The final value of k used for estimating the frame time corresponding to the experiment data is averaged over the number of predecessor sets. With reference to Figure 3.7, the following are obtained:

1. The moving average of the gradient with a window of one frame is 7.1423×10^{-4}. The estimated frame rate at a steady-state vertex count

input of 120,000 is 85.7076 FPS. However, the measured frame rate is approximately 104. Hence the error is approximately 18.3 FPS.

2. The moving average of the gradient with a window of 200 frames is 7.0482×10^{-4}. The estimated frame rate at a steady-state vertex count input of 120,000 is 84.578. However, the measured frame rate is approximately 104. Hence the error is approximately 19.4 FPS.

3. The moving average of the gradient with a window of 500 frames is 6.6301×10^{-4}. The estimated frame rate at a steady-state vertex count input of 120,000 is 79.56. However, the measured frame rate is approximately 104. Hence the error is approximately 24.44 FPS.

The above results obtained from the second frame time estimation technique show errors approximately four to six times larger than the output from the system model using our proposed approach. In brief, our modelling framework out-performs both the first and second estimation techniques.

3.8   SUPERPOSITION IN 3D RENDERING SYSTEM MODEL

The system models derived in the previous sections are based on a specific configuration of the rendering state machine. In this section, we want to further investigate and extend the proposition of a system model for the rendering process that may be broken down further into multiple system models. In the context of real-time rendering, this may be explained as the dissection of a rendering process into its constituent components. Why is this important? The formulation of a rendering process system model if proven to adhere to the principle of superposition is pivotal for gaining the following benefits:

- The output of a combination of rendering processes can be determined without additional modelling.
- A suitable controller can be designed for each constituent rendering process model. This provides greater flexibility and accuracy in controlling the output of the combined rendering process.

At this juncture, we want to establish the validity that each constituent process system model contributes to the combined rendering system model. A hypothesis in componentised modelling of 3D rendering based on the principle of superposition is proposed.

3.8.1   PRINCIPLE OF SUPERPOSITION

In system theory, the net response at a given place and time caused by two or more stimuli for linear systems is the sum of the responses that would have been caused by each stimulus individually. Thus, if input A produces response X and input B produces response Y, input $(A + B)$ produces response $(X + Y)$. Mathematically, for all linear systems, $y = F(x)$ where x is some sort of stimulus (input) and y is some sort of response (output), the superposition of stimuli yields a superposition of the respective responses such that:

$$F(x_1 + x_2 + \ldots + x_3) = F(x_1) + F(x_2) + \ldots + F(x_n) \tag{3.18}$$

Using the same principle, we propose that the overall rendering function of an application is equivalent to the sum of the individual functions of the batch jobs lined in the render queue. To further illustrate, consider a 3D scene with n 3D objects, each with polygon count x_n. The total number of polygons X would be:

$$X = \sum_{m=1}^{n} x_m \tag{3.19}$$

Based on the principle of superposition in Equation (3.19), we draw the parallel analogy that the time taken to render all objects in a scene is equivalent to the sum of the time taken to render each of the individual 3D objects in the scene as given by the following equation:

$$F(X) = \sum_{m=1}^{n} f(x_m) \tag{3.20}$$

where F is the system model of the parent rendering process and f denotes the system model of the separate rendering processes, all obtained through black-box modelling. The assumptions associated with this hypothesis are:

- State changes and context switch overheads between rendering the 3D objects are negligible.
- All objects render within the linear range of the rendering model of the application.
- The application's rendering process is largely partitioned by its content as well.

3.8.2 EXPERIMENT

To validate our hypothesis, we designed an experiment. A 3D rendering application able to display multiple and different types of objects in a single scene was selected. Each type of object was to be rendered in a different way and the number of objects of a type were to be changed during runtime by a user-specified variable. At any time during a run, one or more categories of objects could be rendered.

The application was first set to run with display of only a certain type of object. A data set was defined to consist of a frame rate (output) and total number of objects/triangle count (input) of a type of object. A series of data sets were collected over various object counts within a certain range allowed in the application. The purpose of this step was to collect data so that the rendering process involving one type of object could be modelled.

Step 2 was run again with another type of object and the same types of data were collected. This process was repeated for all types of objects in the 3D scene to enable us to derive the system models for rendering all types of objects separately.

Finally, the application was run with all types of objects displayed and data sets collected by varying the object counts for all the objects. The purpose of this step was modelling the full application so that the overall system model could be compared with the sum of the individual system models obtained in steps 1 and 2.

3.8.2.1 Test Application

A test application was adopted from the NVIDIA DirectX 9.5 SDK. This sample application demonstrated rendering using the hardware instancing technique. The 3D scene in this application consisted of three types of objects (rocks only, spaceships only, and both rocks and spaceships as shown in Figure 3.18). The test application allowed the user to switch off the rendering for any type of object and change the number of objects (for each type) to be rendered as well.

Our data collection procedure started by setting the application to render only one type of object (rocks). The triangle count (input) and frame rate (output) data pairs were collected over multiple object counts within the allowed range. The application then ran with only spaceships displayed. The same data pairs were then collected for a range of object counts. Subsequently, the application was run with both types of objects and the same data collection process. The intent was to constrain the rendering process to specific types of objects so that we could perform black-box modelling to develop the respective system models.

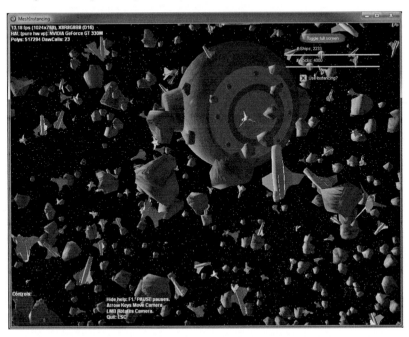

FIGURE 3.18 (See colour insert.) Screenshot of test application in superposition experiment.

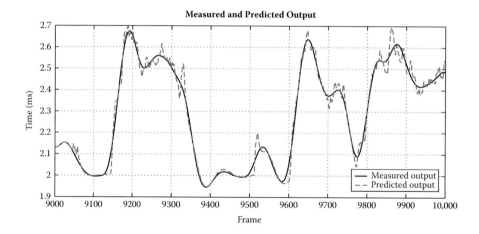

FIGURE 3.19   Measured output and predicted output from Model A.

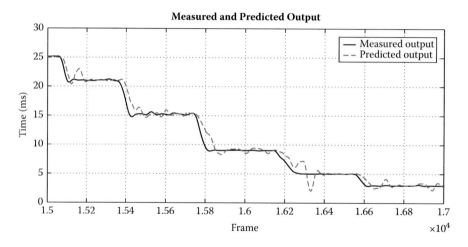

FIGURE 3.20   Measured output and predicted output from Model B.

3.8.3   SIMULATION

By using the MATLAB system identification toolbox and the collected data, we generated the following ARX models for the rendering process:

Model A: only rocks
Model B: only spaceships
Model C: both rocks and spaceships

Figure 3.19 compares the outputs of the measured and system models from the rendering process for rocks only. Figure 3.20 presents the same comparison of rendering only for spaceships. Finally, the output of Model C is compared with the measured data in Figure 3.21.

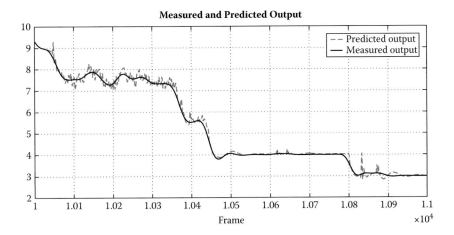

FIGURE 3.21 Measured output and predicted output from Model C.

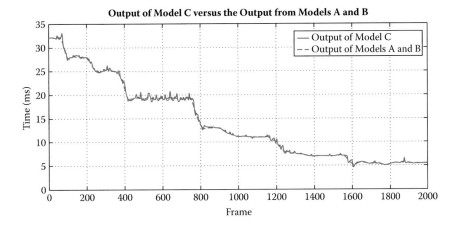

FIGURE 3.22 Comparison of outputs from Model C and summed outputs of Models A and B.

We can observe from these figures that the outputs from the derived system models match closely with the actual measured outputs from the rendering process. Model A's output has a mean error of less than 0.2 ms over a validation range of 10,000 frames. Model B's output mean error is approximately 3 ms for 17,000 frames. Figure 3.21 indicates that the mean error from Model C's output is approximately 1 ms over 11,000 frames which may be considered very low.

Recall from Section 3.8.1 the description of the principle of superposition based on Equations (3.18) and (3.19). The accuracy of the derived system models allowed us to proceed further with validating the principle of superposition by comparing the sum of the individual output of Models A and B with the output of Model C as shown in Equation (3.18) based on the same input data.

From Figure 3.22, we can see that the outputs from Models A and B follow that of Model C very closely. From the experiment data, the peak error is less than 3 ms and

TABLE 3.6
Parameters of ARX Model A in Superposition Experiment

Parameter	Calculation
A(q)	$1 - 4.984 (+ -0.001374) q^{-1} + 9.943 (+ -0.005488) q^{-2} - 9.925 (+ -0.008229) q^{-3}$ $+ 4.958 (+ -0.005488) q^{-4} - 0.9914 (+ -0.001374) q^{-5}$
B(q)	$9.906 \times 10^{-14} (+ -1.729 \times 10^{-13}) q^{-1}$
Operating point	$u = 1.1526 \times 10^5, y = 22.0855$

TABLE 3.7
Parameters of ARX Model B in Superposition Experiment

Parameter	Calculation
A(q)	$1 - 3.008 (+ -0.000723) q^{-1} + 3.023 (+ -0.001451) q^{-2} - 1.015 (+ -0.0007297) q^{-3}$
B(q)	$0.0001093 (+ -4.514 \times 10^{-6}) q^{-8} - 0.0004042 (+ -1.786 \times 10^{-5}) q^{-9}$ $+ 0.0005572 (+ -2.66 \times 10^{-5}) q^{-10} - 0.0003383 (+ -1.767 \times 10^{-5}) q^{-11}$ $+ 7.605 \times 10^{-5} (+ -4.42 \times 10^{-6}) q^{-12}$
Operating point	$u = 2.3494 \times 10^5, y = 17.5628$

TABLE 3.8
Parameters of ARX Model C in Superposition Experiment

Parameter	Calculation
A(q)	$1 - 6.471 (+ -0.007594) q^{-1} + 18.07 (+ -0.04455) q^{-2} - 28.26 (+ -0.1099) q^{-3}$ $+ 26.77 (+ -0.1459) q^{-4} - 15.38 (+ -0.11) q^{-5} + 4.969 (+ -0.04464) q^{-6}$ $- 0.6968 (+ -0.007617) q^{-7}$
B(q)	$1.233 \times 10^{-6} (+ -3.072 \times 10^{-7}) q^{-9} - 6.191 \times 10^{-6} (+ -1.516 \times 10^{-6}) q^{-10}$ $+ 1.25 \times 10^{-5} (+ -3.01 \times 10^{-6}) q^{-11} - 1.269 \times 10^{-5} (+ -3.006 \times 10^{-6}) q^{-12}$ $+ 6.483 \times 10^{-6} (+ -1.51 \times 10^{-6}) q^{-13} - 1.332 \times 10^{-6} (+ -3.05 \times 10^{-7}) q^{-14}$
Operating point	$u = 1.6427 \times 10^5, y = 25.2834$

the mean error is approximately 1 ms for 2,000 frames. The parameters of the three rendering application system models are presented in Tables 3.6 to 3.8.

3.8.4 SUMMARY

In this section, we proposed the compliance of the rendering process to the principle of superposition and validated this hypothesis systematically via experiments. This investigation leads to the conclusion that the summation of separate rendering process outputs (frame rates) is equivalent to the output of a single system model using the combined inputs. In terms of research significance, an accurate system model can be built upon by the concatenation of separate constituent rendering processes. This is particularly useful when devising a system model for a complicated rendering process is difficult. Furthermore, this principle provides a user with additional flexibility to manipulate application rendering at a higher resolution.

3.8.5 ADDITIONAL NOTES

With reference to Equations (3.18) and (3.19), it is important to note that the mathematical representation of the system model f is not unique even though it produces the same stable state output given a same set of input and rendering states. This is because the dynamics of the rendering system will vary at different operating (input and output) ranges.

3.9 CONCLUSION

In this chapter, we demonstrated in a systematic manner how our proposed data-driven modelling framework can produce accurate linear models of real-time rendering for a variety of applications. We illustrated the extensibility of our framework to handle multiple inputs and validated the accuracy of the resultant model. More importantly, we validated the case by which the models produced by such a framework are ultimately useful in the context of interactive rendering with the introduction of a controller. Finally, our control system is able to eliminate the frame rate oscillation problem found in typical reactive scheduling frameworks.

Our framework is designed to work on polygonal-based rendering pipelines found in commodity graphics hardware and it leverages geometry subdivision as a primary basis for process modelling. As a future research endeavour, we will try to expand the scope of the model variables for various types of rendering processes wherever appropriate and possible.

At this juncture, our work is largely based on the subdivision of a single large mesh. This is useful for applications involving a single object of interest such as massive model rendering and computer-aided design. However it can be extended to support multiple progressive meshes in more elaborate applications such as games.

4 Modelling Non-Linear Rendering Processes

4.1 INTRODUCTION

The real-time rendering process is inherently non-linear [38]. This can be understood from the fact that computer systems on which software runs are constructed using electronic components that exhibit non-linear material properties. Consequently, using a single linear model to describe the behaviour of a non-linear system would be inadequate. In this chapter, we describe an approach by which this non-linear characteristic can be captured sufficiently with appropriate system models using advanced techniques in soft computing.

4.2 BACKGROUND

4.2.1 System Modelling with Neural Networks

In system identification, it is often necessary to begin with the assumption that the underlying model is linear and then apply the appropriate model structures described above. However, an actual system may not always exhibit linear characteristics throughout an operating range. For example, research conducted by Hook and Bigos [38] showed that the time required to process a single vertex varies even when parameters such as rendering states and display resolution are fixed. Therefore, it is useful to conduct a comprehensive analysis to better understand the dynamics of a system.

In this research, we introduce the application of artificial neural networks (ANNs) to model the non-linearity in the real-time rendering process. Soft computing techniques based on the artificial neuron proposed by McCulloch and Pitts [39] spread widely into many other fields of study in recent decades. The distinctive nature of the artificial neurons in various network configurations provided the capability to model both linear and non-linear systems with good accuracy.

The first artificial neuron proposed by McCulloch and Pitts mimicked the functioning of biological neurons through a multiple-input–single-output model. The artificial neuron is essentially a processing unit that sums the weighted values of its inputs to produce an intermediate output that is then fed as an input to an activation function that produces the final output. An ANN is formed with layers of interconnected neurons and is frequently used to simulate the functions of many systems.

Figure 4.1(a) illustrates the structure of the artificial neuron. ANNs must be trained to capture the characteristics of the systems they model. Training algorithms

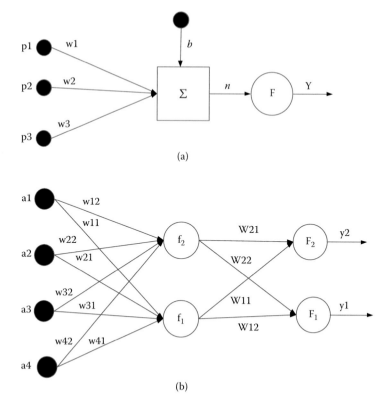

(a)

(b)

FIGURE 4.1   (a) Perceptron neuron. (b) Multi-layer perceptron network (MLP).

for neural networks such as the back-propagation [40] and Levenberg-Marquardt [41] methods were developed to compute the weights and biases for inputs.

We adopted neural networks in our research for modelling the rendering process because of their ability to capture information from complex, non-linear, multi-variate systems without the need to assume underlying data distribution or mathematical models. In recent years, the popularity of using multi-layer perception networks has increased due to their successes in real-world applications such as pattern recognition and control applications.

Dynamic neural networks use memory and recurrent feedback connections to capture temporal patterns in data. Waibel et al. [42] introduced the distributed time delay neural network (DTDNN) for phoneme recognition. An extension of this network structure gives the flexibility to have tapped delay line memory at the input to the first layer of a static feed-forward network and throughout the network as well. For general discussion, a two-layer DTDNN is presented in Figure 4.2.

The choice of using ANNs to model a computing process such as real-time rendering may be explained easily. First, the dynamics of an ANN arising from delay units within its structure provides an inferred correspondence with the architecture of current computing hardware. To illustrate, a delay usually occurs in embedded circuits when data are transferred between the processor and memory units.

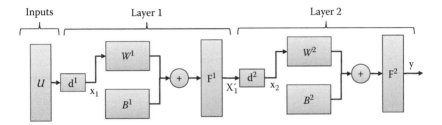

FIGURE 4.2 Two-layer distributed time delay neural network with time delays at inputs of each layer. The notations with their respective meaning or representative are:

U, the input layer
d, the delay
Wn, where *W* is the weight and *n* represents the *n*th layer
Bn where *B* is the bias and *n* represents the *n*th layer
Fn, where *F* is the firing function and *n* represents the *n*th layer
y, the output of the network

Second, the ANN's layered network structure makes it easy to extend by cascading ANNs together for modelling modular systems. For example, to model a complicated system it may be possible to break down the process into modelling individual subsystems using simple ANNs and then joining them together. This is certainly applicable in the context of computer software since modern programming paradigms emphasise modularity and object-oriented principles.

4.2.2 Systems Modelling with Fuzzy Logic

Fuzzy set theory and fuzzy control have been implemented successfully in many technical fields. The primary benefit offered by the fuzzy control paradigm is its ability to emulate human control based on linguistic variables and a set of intuitive expert rules used as a decision or inference system. In comparison to conventional control techniques, the advantages of the fuzzy control paradigm are twofold.

First, it imposes no requirement for a mathematical model of the system to be controlled. This is especially important and useful as it may be difficult to derive certain process models due to their complex dynamics and some systems cannot be modelled using first principles. Second, the fuzzy controller works on relatively straightforward computation and can be developed to handle non-linear processes empirically in practice without the need for complicated mathematics.

In addition, fuzzy logic is tolerant of imprecise data. Systems with reliable performance can be built using fuzzy logic that leverages the experiences of experts. In direct contrast to neural networks that use training data and generate system models, fuzzy logic allows a user to rely on the experiences of humans who understand the system.

Furthermore, fuzzy logic can be blended with conventional control techniques. In many cases, fuzzy systems augment them and simplify other implementations. Finally, fuzzy logic is based on natural language that provides a strong basis for human communication. As a result, fuzzy logic is easy to use. These advantages translate to its appeal as a practical solution to real world control problems involving implementation.

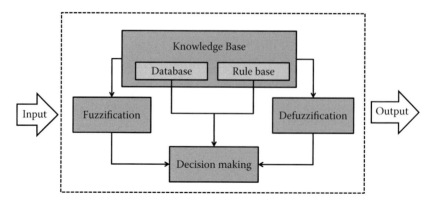

FIGURE 4.3 (See colour insert.) Fuzzy inference system.

In general, a fuzzy inference system (Figure 4.3) consists of five functional components:

1. A fuzzification process that transforms discrete values (inputs) into various degrees of membership with linguistic values
2. A rule base containing a set of fuzzy if–then rules
3. A set of membership functions of the fuzzy sets used in the rule base
4. A decision-making process that performs fuzzy inference operations on the rules
5. A defuzzification process that maps fuzzy results from the inference engine to a crisp output

The process for fuzzy reasoning performed by a fuzzy inference system is as follows.

1. *Fuzzify* the input values by comparing the input variable with the membership function to obtain their corresponding membership values.
2. Combine the membership values of all the premise components to find the firing strength of each rule.
3. Generate the consequent results from each rule depending on the firing strength.
4. *Defuzzify* the results by aggregating the qualified consequents to produce the final crisp value.

The development of a fuzzy control system begins with the two key components: (1) the input–output membership functions describing the properties of the system (fuzzy sets) based on linguistic variables and (2) the rule-base that relates the input–output sets. Given an antecedent and consequent relationship between an input y to a SISO system's output u using linguistic descriptions of their properties, the calculation may be represented as

$$IF \ y \in Y_j \ THEN \ u \in U_j \tag{4.1}$$

In each universe of discourse, U_i and Y_i and u_i and y_i take on values with corresponding linguistic variables $\widetilde{u}_i$ and $\widetilde{y}_i$ that describe the characteristics of the variables. Suppose A_i^j denotes the jth linguistic value of the $\widetilde{u}_i$ linguistic variable defined over the universe of discourse U_i. If we assume that many linguistic values are defined in U_i, the linguistic variable $\widetilde{u}_i$ that takes on the elements from the set of linguistic values may be denoted by Equation (4.2).

$$\widetilde{A}_i = \left\{ \widetilde{A_i^j} : j = 1,2,\ldots,N_i \right\} \tag{4.2}$$

In the same manner, we can consider that $\widetilde{B_i^j}$ to denote the jth value of the linguistic variable $\widetilde{y}_i$ defined over the universe of discourse Y_i. $\widetilde{y}_i$ may be represented by elements taken from the set of linguistic values denoted by the following equation.

$$\widetilde{B}_i = \left\{ \widetilde{B_i^p} : p = 1,2,\ldots,M_i \right\} \tag{4.3}$$

Given a condition where all the premise terms are used in every rule and a rule is formed for each possible combination of premise elements, we have rule set with N_i number of rules that can be expressed as:

$$\prod_{i=1}^{n} N_i = N_1 \cdot N_2 \cdot \ldots \cdot N_n \tag{4.4}$$

Based on the membership functions, the conversion of a crisp input value into its corresponding fuzzy value is known as fuzzification. The defuzzification of the resultant fuzzy set from the inference system to a quantifiable value may be done using the centroid (centre of gravity) method [43]. The principle is to select the value in the resultant fuzzy set such that it would lead to the smallest error on average given any criterion. To determine y^*, the least square method can be used and the square of the error is accompanied by the weight of the grade of the membership function $\mu_B(u)$. Therefore, the defuzzified output y^* may be obtained by finding a solution to the following equation.

$$y^* = \arg\min_{y^*} \int \mu_B(y)\left(y^* - y\right)^2 du \tag{4.5}$$

Differentiating with respect to y^* and equating the derivative to zero yields:

$$y^* = \frac{\displaystyle\int_Y y\mu_B(y)dy}{\displaystyle\int_Y \mu_B(y)dy} \tag{4.6}$$

which gives the value of the abscissa of the centre of gravity of the area below the membership function $\mu_B(u)$.

4.3 EXPERIMENTS

In this section, we describe the experiments conducted to model two real-time rendering applications. The approaches are premised upon the neural network and fuzzy modelling techniques mentioned in Section 4.2. In all experiments, empirical data consisting of the per-frame triangle count and frame rate were collected from the two different applications running on a Pentium IV, 3.2 GHz processor with 2 GB RAM and NVIDIA's GeForce 6800 graphics board.

In the data collection process, the user is free to move the camera view to simulate common navigation patterns or object manipulation in virtual environments. This action is designed so that a wide range of polygon loads and a good combination of rendering features may be captured. All applications rendered the animated frames in real time according to the input of the user.

4.3.1 Time Delay Neural Network

To illustrate the applicability of time delay neural networks in modelling the rendering process, we selected two applications with different levels of complexity. The first application was developed to encompass most common rendering parameters in applications such as textures, fog, lighting, animation, shader effects, and moderate depth complexity. It consisted of a scene populated by hundreds of instances of a 3D object (a virtual character with a certain surface shading effect) appearing with an animated landscape. A screenshot of this application is provided in Figure 4.4.

In contrast to the more controlled environment in the first experiment, the application in the second experiment was taken from a popular game software system called "Serious Sam 2"© (2KGames, www.croteam.com). The test case was selected for its complex rendering functions and scene composition. Figure 4.5 is a screenshot of this software.

In the second experiment, a certain game environment was selected based on the level of complexity and the rendering statistics were collected. To capture the low-level data used in the real-time rendering processes, we used Microsoft's DirectX tool, PIX Performance Analyzer [44], and utilities from NVIDIA's NVPerfKit [45]. The MATLAB® Neural Network Plant Identification Tool [46] was utilised for modelling the rendering process.

In accordance to the system identification methodology described in Chapter 3, a neural network was first selected as the model structure. The collected data were fed into the neural network to train it to generate an accurate mapping of the relationship between the input triangle count and the output frame rate. Different neural network structures and parameters were tested to determine the best fitting model. This process continued iteratively until the performance objective (a numeric quantity describing the difference between the predicted and actual frame rates) was met. The same procedure was repeated for both experiments.

4.3.2 Adaptive Neuro-Fuzzy Inference System (ANFIS)

In addition to neural networks, we introduced the concept of using fuzzy system modelling for real-time rendering in Section 4.2.2. In Experiment 3, we adopted the adaptive neuro-fuzzy inference system (ANFIS) to achieve this objective.

FIGURE 4.4   **(See colour insert.)** Screenshot of application in Experiment 1.

FIGURE 4.5   **(See colour insert.)** Screenshot of application in Experiment 2.

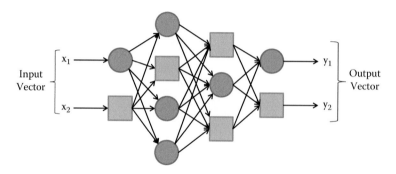

FIGURE 4.6 Adaptive network.

The adaptive neuro-fuzzy inference system was introduced by Jang [47]. It is essentially a fuzzy inference system implemented in the framework of adaptive networks. The proposed architecture utilises a learning procedure of the adaptive network which is a superset of general feed-forward neural networks with supervised learning capability.

Based on this adaptive design, the outputs of the framework depend on parameters pertaining to the nodes involved and the learning rule specifies how the parameters should change to minimise a prescribed error metric. The back-propagation algorithm or least square method may be used in such computation. The relative advantage is that this technique is capable of automatically constructing an input–output mapping based on both human knowledge and experimental data. Figure 4.6 presents the design of a basic ANFIS.

The mathematics behind the ANFIS architecture is described as follows. First, we assume a given adaptive network with L layers and k nodes in the Kth layer. We use the notation (k, i) to describe the node at the ith position of the Kth layer with its node function O_i^k. In neural networks, node output is determined by the input signals and the note parameter set. Hence we denote this output by y_i^k:

$$y_i^k = O_i^k \left(O_i^{k-1}, \ldots, O_{k-1}^{k-1}, a, b, c \ldots \right) \tag{4.7}$$

where a, b, c... are the parameters of this node. Next, assuming the given data has P entries, the error metric for the pth entry may be defined as the sum of squared errors:

$$E_p = \sum_{m=1}^{L} \left(T_{m,p} - O_{m,p}^L \right)^2 \tag{4.8}$$

$T_{m,p}$ is the mth component of the pth target output vector, and $O_{m,p}^L$ is the mth component of the actual output vector produced by the presentation of the pth input vector. The overall error measure is given by:

$$E = \sum_{p=1}^{P} E_p \tag{4.9}$$

The learning procedure using gradient descent over the parameter space requires error rates to be computed for the pth training data and for each node's output O given by:

$$\frac{\partial E_p}{\partial O^L_{i,p}} = -2\left(T_{i,p} - O^L_{i,p}\right) \quad (4.10)$$

The error rate for the internal node at (k, i) can be derived using the chain rule:

$$\frac{\partial E_p}{\partial O^k_{i,p}} = \sum_{m=1}^{k+1} \frac{\partial E_p}{\partial O^{k+1}_{m,p}} \frac{\partial O^{k+1}_{m,p}}{\partial O^k_{i,p}} \quad (4.11)$$

where $1 \le k \le L - 1$. Given α as a parameter of the given adaptive network, we have

$$\frac{\partial E_p}{\partial \alpha} = \sum_{O^* \in S} \frac{\partial E_p}{\partial O^*} \frac{\partial O^*}{\partial \alpha} \quad (4.12)$$

where S is the set of nodes whose outputs depend on α. We can get the derivative of the overall error measure E with respect to α with Equation 4.13.

$$\frac{\partial E_p}{\partial \alpha} = \sum_{p=1}^{P} \frac{\partial E_p}{\partial \alpha} \quad (4.13)$$

Furthermore, we can describe the update formula for α as Equation 4.14.

$$\Delta \alpha = -n \frac{\partial E}{\partial \alpha} \quad (4.14)$$

in which n is the learning rate.

Equations (4.6 to 4.14) describe the structure and learning process of the adaptive network. In an ANFIS architecture, this network should be functionally equivalent to a fuzzy inference system. To illustrate this mapping, consider a simple case of an ANFIS system with two inputs x_1 and x_2 and one output, y. Suppose the rule-base contains two fuzzy *IF-THEN* rules. Then we may write

Rule 1: *IF x_1 is A_1 and x_2 is B_1, THEN $f_1 = p_1 x_1 + q_1 x_2 + r_1$*
Rule 2: *IF x_1 is A_2 and x_2 is B_2, THEN $f_2 = p_2 x_1 + q_2 x_2 + r_2$*

where A and B are antecedents and f is the output of the neuron (node) in the same layer, p, q and r are the parameters specific to the node. In the adaptive network, the membership function describing an antecedent can be denoted by the following node function.

$$O^1_i = \mu_{A_i}(x) \quad (4.15)$$

where x is the input to the node i, A the linguistic label (antecedent) associated with this node function. In terms of the choice of membership function characteristics, Jang [47] proposes the typical bell-shaped function, which is adopted in this research and found to be adequate with minor adjustments.

For practical applications, the modelling approach using ANFIS is similar to many system identification techniques. First, a hypothetical parameterised model structure that relates the inputs to membership functions to rules to outputs to membership functions is selected. Thereafter, a set of input-output data collected from an experiment is used for the ANFIS training. A portion of the same set of data is reserved for validation of the derived system model. In an iterative manner, the FIS model can be trained to emulate the data presented to it by modifying the membership function parameters according to a chosen error criterion.

4.4   EXPERIMENT RESULTS

4.4.1   Time Delay Neural Networks

The neural network used to model the first application consisted of a MLP network with two layers, six units, and three delay units in each of the input and output channels. The second neural network differed and contained just four delay units in both the input and output channels. The neural network used to model the first application is shown in Figure 4.7.

In Figure 4.8, the diagrams at top right and bottom right show the measured and predicted output frame rates of the application and neural network, respectively. The difference between them is shown in the graph at bottom left. The graph in the top left corner shows the input (triangle count per frame) to the neural network model

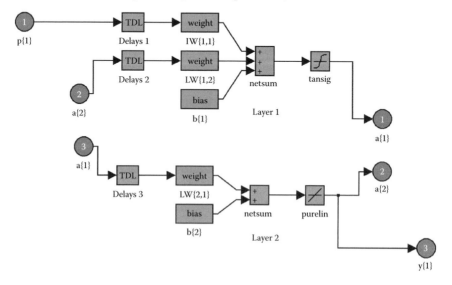

FIGURE 4.7   Neural network in Experiment 1.

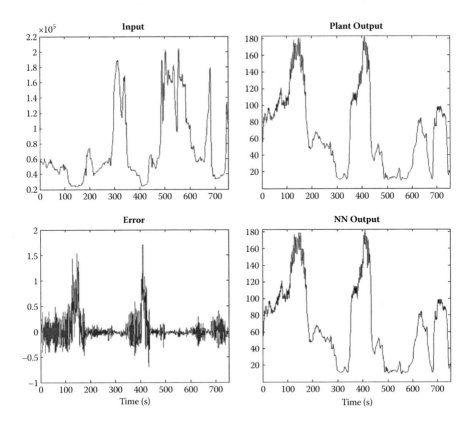

FIGURE 4.8   Data collected from Experiment 1.

over the test period. The mean difference between the frame rates generated by the neural network model and the actual application was 0.00455.

In the same order, the graphs for the second experiment using the game are shown in Figure 4.9. The neural network was able to model closely the characteristics of the rendering process in the second application with a mean difference of 0.00896 in frame rate. All networks were trained using the Levenberg-Marquardt algorithm over 200 epochs for over 5,000 frame samples.

4.4.2   ANFIS MODEL

In Experiment 3, 120,000 input and output data pairs, each consisting of a vertex count and frame rate, were collected. Figure 4.10 is a screenshot of the 3D rendering application. Eighty thousand data pairs were used for training the ANFIS and the remaining data pairs for validation. We used the ANFIS tool from the MATLAB Fuzzy Logic Toolbox for the design and training of the fuzzy inference system.

The ANFIS model output was compared with the user-defined reference output in Figure 4.11. We can observe from the figure that the output of the ANFIS model closely follows the reference output. The error over the entire duration of the

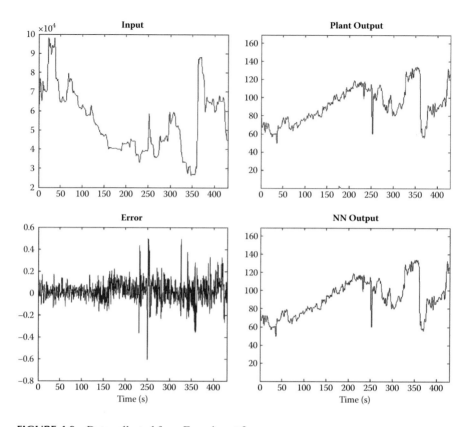

FIGURE 4.9 Data collected from Experiment 2.

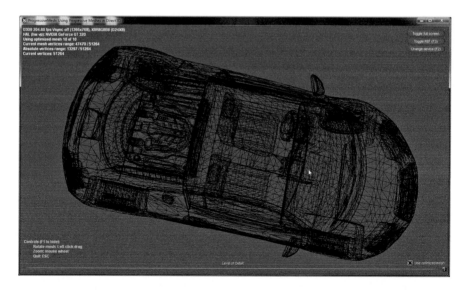

FIGURE 4.10 (**See colour insert.**) Screenshot of rendering application in Experiment 3.

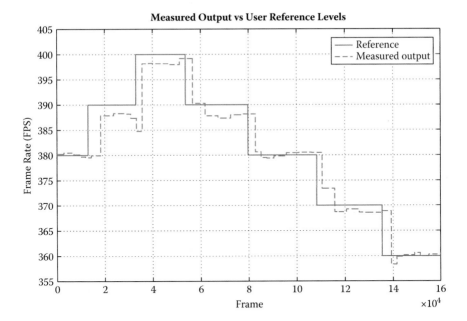

FIGURE 4.11 Measured and reference output from ANFIS in Experiment 3.

experiment was less than 5 FPS. A delay registered in the experiment results may be attributed to the latency between the connected peripherals in the experiment setup.

4.5 DISCUSSION

The common objective of our tests was to derive an accurate system model of the rendering processes. While it may seem ideal to have a single model for all applications, a single model is impractical because various rendering applications have different dynamics and vary in the numbers of components contributing to the final render time.

For example, applications differ in the types and numbers of processes such as network communication, application logic, and input–output computations. Hence it is not a trivial task to derive a universal model for all rendering applications. Furthermore, a generalised model would not necessarily be useful because it might not provide a user with a set of components that could be used easily in the rendering process.

Another benefit from using soft computing techniques such as neural networks and fuzzy systems is that they provide greater convenience for modelling wider operating ranges compared to using linear model structures. They eliminate the need to conduct several tedious data collection procedures over an operating range.

Furthermore, when a satisfactory model is derived, there is no need to re-train the neural network or ANFIS unless the construct of the application changes. As to speed of modelling, the training of our neural networks typically required fewer than 3 minutes for a dataset of approximately 5,000 data points on a mid-end desktop computer.

Finally, we provide the mathematical basis by which even linear models can be derived from the non-linear models obtained from empirical data in Section 4.6.

4.6   LINEARISED APPROXIMATION FROM NON-LINEAR MODELS

This section introduces linearisation of a non-linear system model (at a particular operating point) such as the ANN described previously. The need to extract linear properties of a non-linear model often arises because many systems function largely in a specific range instead of spanning an entire operating range. Furthermore, it is easier to work with linear systems due to the mathematics involved. With reference to Figure 4.7, it is possible to envision how a non-linear function may be approximated by a series of linear segments over different ranges.

We provide the mathematical derivation for linearisation of a DTDNN below. The linearised model takes the form of state space [8] equations that are common in many system identification and control studies. In addition to their maturity in the field of mathematics, state space equations provide mathematical constructs that leverage linear, first-order derivative variables that are convenient for both computation and extension. For linear systems, the state space equations are:

$$x(k+1) = Ax(k) + Bu(k) + Ke(k) \tag{4.16}$$

$$y(k) = Cx(k) + Du(k) \tag{4.17}$$

where $x(k)$ is the state vector, $y(k)$ is the system output, $u(k)$ the system input, and $e(k)$ the stochastic error. A, B, C, D, and K are the system matrices.

Equations (4.16) and (4.17) describe the relationship of the internal states, input, and output of the system. The state variables are denoted by x_1 and x_2 while the input and output of the neural network are u and y, respectively. W_i denotes the weights assigned at the neurons on layer i while B_i refers to the corresponding bias value on the same layer. The triggering function at each layer of the neural network is denoted by F_i, which typically may be linear, sigmoid, or threshold in nature.

Unit time delays were introduced at the input stage of each layer, as denoted by the d_i blocks. Since the time delays relate to the dynamics of the neural network, the related equations are presented with a time step variable k that indicates its correspondence in terms of implementation in digital systems such as computers.

$$x_1(k+1) = u(k) \tag{4.18}$$

$$x_2(k+1) = F^1\left(W^1 x_1(k) + B^1\right) \tag{4.19}$$

$$y(k) = F^2\left(W^2 x_2(k) + B^2\right) \tag{4.20}$$

The linearised approximation of the model at an operating (trim) point is:

$$\Delta x(k+1) = A\Delta x(k) + B\Delta u(k) \tag{4.21}$$

$$\Delta y(k) = C\Delta x(k) + D\Delta u(k) \tag{4.22}$$

where Δx_1, Δx_2, Δu, and Δy are small deviations:

$$\Delta x_1(k) = x_1(k) - x_{1trim} \tag{4.23}$$

$$\Delta x_2(k) = x_2(k) - x_{2trim} \tag{4.24}$$

$$\Delta u(k) = u(k) - u_{trim} \tag{4.25}$$

$$\Delta y(k) = y(k) - y_{trim} \tag{4.26}$$

with

$$x_{1\,trim} = u_{trim} \tag{4.27}$$

$$x_{2\,trim} = F^1\left(W^1 u_{trim} + B^1\right) \tag{4.28}$$

$$y_{trim} = F^2\left(W^2 x_{2\,trim} + B^2\right) \tag{4.29}$$

$$C = \frac{\partial}{\partial x} F^2(x,u) \;|_{x_{trim},y_{trim}} \tag{4.30}$$

$$D = \frac{\partial}{\partial u} F^2(x,u) \;|_{x_{trim},u_{trim}} \tag{4.31}$$

From the above equations, we have:

$$\Delta y(k) = y(k) - y_{trim} = \frac{d}{dx_2}\left\{F^2(W^2 x_2(k) + B^2)\right\} \tag{4.32}$$

and

$$A = \begin{bmatrix} 0 & 0 \\ \dfrac{d}{dx_1}\left\{F^1(W^1 x_{1trim} + B^1)\right\} & 0 \end{bmatrix}, \; B = \begin{bmatrix} 1 \\ 0 \end{bmatrix} \tag{4.33}$$

This linearised model representation of the rendering process makes it possible for a user to design a control system in which the output of this system model can be driven to produce the stable frame rates required in interactive applications.

4.7   CONCLUSION

This chapter described approaches to modelling the non-linear real-time rendering process. Since linear models cannot fully capture the characteristics of certain rendering processes and typically cover a larger operating range, we proposed the use of neural networks and the fuzzy inference system. The application of these techniques was demonstrated in experiments with various rendering processes. The results indicate that both techniques are capable of producing accurate system models from the measured data.

5 Model-Based Control

5.1 INTRODUCTION

Research in real-time computer graphics focuses on trading speed of rendering for image quality but does not address the problem of frame rate stability—a critical component of the user experience. Common techniques offer "best effort" solutions to achieve interactive frame rates without any performance guarantee. Consequently, the onus of finding an optimal solution is left to the application developer if not totally forsaken. In the absence of a feasible solution, investments in many interactive applications such as those from the training, visualisation, and simulation domains may not yield adequate results. Previous research [48,49] has shown the importance of maintaining interactive frames in these applications.

Control engineering is a mature field of study with myriad applications in various systems that affect our daily lives. Its efficiency when applied to electrical and mechanical systems in fields as varied as aerospace, defence, communications, and manufacturing equipment has been proven in numerous industries [50,51] around the world. Little research relates the adaptation of control theory to real-time computer graphics rendering. However, in recent years, we noted observable momentum of cross-disciplinary research in control theory and computer systems [2,7,52,54].

In this chapter, we introduce the concepts of control theory and demonstrate relevant techniques as mechanisms for achieving sustainable performance in real-time computer graphics rendering.

5.2 CONTROL SYSTEM PERSPECTIVE OF COMPUTER GRAPHICS RENDERING PROCESS

We consider the computing environment for real-time rendering to consist of a homogeneous infrastructure consisting of both hardware and software. A simple representation of the rendering system is shown in Figure 5.1.

The rendering process is modelled as the *plant* in control taxonomy. Its basic functionality is to process a stream of inputs such as 3D geometry and other rendering data to create a series of images in real time. Other processes running in the same computing environment may periodically share the memory and CPU time thereby creating interruptions that may be represented as *disturbances*. Furthermore, the rendering system can receive input from user interaction with the computing environment.

To meet the goal of consistent and sustainable frame rates from the output of the plant, we introduce a *controller* as an extension of the system shown in Figure 5.1.

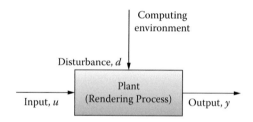

FIGURE 5.1 **(See colour insert.)** Rendering process from system perspective.

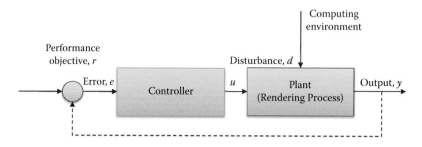

FIGURE 5.2 **(See colour insert.)** Closed-loop feedback control system.

Figure 5.2 depicts a feedback control system in which this controller is adopted to regulate input to the plant to achieve a certain performance objective.

In this configuration, the rendering process produces a series of images, each in a certain amount of time (frame time) and this plant output is compared with a pre-defined performance metric through a data feedback channel for every cycle of a rendered frame. To make the comparison of the output of the plant and the performance objective useful, these two data streams must be expressed in the same unit of measurement. Typically, the time taken to render one frame of image or its mathematical inverse (frame rate) is the measurement unit. The error between the two quantities is passed to the controller that subsequently generates a control action for the plant.

One interesting point is that the performance objective may be predefined by a user or dynamically set by a more elaborate system that measures quality of service (QoS) in the computing environment. Furthermore, this closed-loop feedback control system provides corrective action even when disturbance from the computing environment occurs.

5.2.1 CONTROL SYSTEM ARCHITECTURES FOR REAL-TIME RENDERING

A prudent and imperative step in control system design is understanding the plant characteristics to be controlled. Real world systems and processes seldom display linear characteristics over their operating ranges because the physical nature of materials used creates non-linearity in integrated systems.

Real-time rendering is also complex because of the numerous inputs and configuration settings. A plant with varying dynamics would justify the use of an adaptive controller to meet system performance objectives. In Figure 5.3, a QoS component

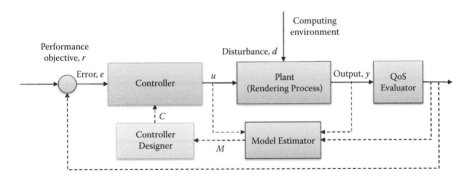

FIGURE 5.3 **(See colour insert.)** Rendering system with adaptive controller and quality of service feedback.

is introduced as an additional evaluation step to compute qualitative performance in addition to evaluating plant output only.

Furthermore, a model estimator (Figure 5.3) may be used to provide periodic assessments of plant dynamics so that an appropriate control strategy can be computed and implemented to meet performance requirements. This approximation of the plant model forms an important basis to allow a designer to make decisions about changing control parameters or introducing new control laws into a control system. Astrom and Wittenmark's research on adaptive control [53] provides insight into controller design based on a plant with uncertain parameters and dynamics.

The advantage of the control system described in Figure 5.3 lies in the flexibility of controller design that is not fixed and whose parameters do not need to be known at design time. While elaborate control system designs may be considered plausible solutions to the frame rate inconsistency problem in computer graphics rendering, they may not always be computationally effective for use in real-time applications due to their complexity.

An intuitive step to circumvent this problem is to design a control system in a modularised manner and treat the plant and control as separate subsystems. This architecture provides greater flexibility for the controller and plant because computing resources are dedicated to each subsystem and any disturbance arising from controller-related computation would not affect plant operation. The feedback data channel and feed-through from the controller to the plant can be achieved by network communication. The design of this modular control system is presented in Figure 5.4.

An additional consideration for the design shown in Figure 5.4 is the data transport overhead arising from the inter-subsystem communication. In typical control engineering applications, this can be modelled analogously as delays from actuators and sensors. These delay components are illustrated as components of the communication channels in Figure 5.4.

In summary, we have provided a systematic and progressive introduction of the control system perspective for the real-time rendering. We also presented a high level overview of the various control system architectures and relevant implementation considerations.

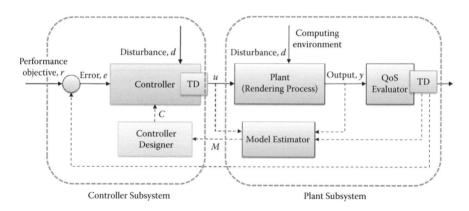

FIGURE 5.4 **(See colour insert.)** Modular adaptive control system for real-time rendering.

5.2.2 CONTROL SYSTEM PERFORMANCE CONCEPTS APPLICABLE TO REAL-TIME RENDERING

The value of incorporating control principles in real-time rendering would be better appreciated by highlighting important concepts pertaining to control system design. These performance objectives are the premises by which the control systems are validated for their effectiveness at the design level. Since exhaustive coverage of this topic is beyond the scope of this chapter, we focus the discussion on certain characteristics.

Stability—A system is inherently stable if it is not easily perturbed by small variations (disturbances) introduced when it is at equilibrium state. Stability is a system property that may be best described by the bounded-input–bounded-output (BIBO) signal processing nomenclature. One example is a rendering system designed to run at a user-defined frame rate when the controller works in a way that small load variations sent to the rendering process do not result in unstable frame rates at the output end.

Controllability—In control theory, it is possible to consider state and output controllability. For brevity and in the context of real-time rendering, we extract the basic underlying concept—the ability to manipulate or steer the output based on an admissible set of rendering inputs within a specific time window.

Observability—Observability and controllability are mathematical terms for the same problem. The observability of a system refers to how well its internal states may be inferred by knowledge of its outputs and inputs. In simpler terms, a system is observable if the behaviour or current values of its states can be determined by analysing its outputs and inputs. Both observability and controllability criteria reinforce a system with clear requirements for stable operation.

Robustness—Not all systems can handle large and unpredictable plant fluctuations. The robustness of a system is its ability to operate under such situations to achieve its objectives or allow its performance to degrade gracefully without catastrophic failure. Apart from resilience to fluctuating operating conditions, a real-time rendering application generally does not incur significant cost or damage to its environment even when it fails.

While real-time rendering may be considered similar to other computing processes such as Web service, database queries, and network communication, it has distinctively different task handling properties. In the work of Hellerstein et al. [54], the tasks of feedback control are piecewise in nature and may be scheduled according to changes in system loading. However, real-time rendering involves a series of interdependent tasks (pipeline stages) that cannot be chosen selectively for processing. Therefore task scheduling algorithms are not applicable to this process because each rendered frame must follow a sequential order to create visual animation effects.

5.3 PID CONTROL AND TUNING

PID controllers [55] have accumulated a long history since the industrial revolution and are known to operate in more than 80% of the world's control systems. The fundamental PID control algorithm works on simple structures and produces good performance without the need for heavy computation. This means that PID controllers are inherently fast and easy to design, operate, and maintain. The PID control action in a closed-loop feedback system takes a parallel mode form as shown in Equation (5.1).

$$u(t) = K_p e(t) + K_i \int_0^t e(\tau)d\tau + K_d \frac{d}{dt}e(t) \qquad (5.1)$$

At the implementation level, a PID controller's discrete form may be expressed as Equation (5.2)

$$u(n) = K_p e(n) + K_i \sum_{k=0}^{n} e(k) + K_d(e(n) - e(n-1)) \qquad (5.2)$$

where

$$K_i = \frac{K_p T}{T_i}, \, K_d = \frac{K_p T_d}{T}$$

where $u(n)$ is the control action and T_p, T_i, and T_d denote the time constants of the proportional, integral, and derivative terms, respectively.

For a PID controller to be effective, we see from Equation (5.2) that the gain values of the controller must be set correctly. The process of determining these parameters is known as controller *tuning*. A comprehensive summary of the techniques for tuning the PID controller is provided in Reference [55]. For our system, the PID controller is tuned using the Robust Response Time Tuning Algorithm from the MATLAB® Control Design Toolbox [56].

5.3.1 IMPLEMENTING PID CONTROL FOR RENDERING PROCESS

With reference to Figure 5.2, the implementation of a PID controller-based rendering system follows the same design with the controller block represented by a PID controller. The plant handles the rendering process. The procedure for obtaining the rendering process model is described in Chapters 3 and 4.

In this section, we discuss the introduction of a PID controller in a simulation environment and also use the controller in the rendering process. Figure 5.5 shows the design of the closed-loop PID control system adopted in this research. At the output, the numerical value of the frame rate is tapped and sent to a comparator that computes the difference between this output and a predefined frame rate. The error data are sent to the PID controller.

With reference to Equation (5.2), the control action is computed based on the $e(n)$ input and the PID controller's internal structure parameters (gain values). The control action generated by the controller regulates the input to the plant such that the frame rate eventually tracks the predefined target.

As mentioned in Section 5.2, it is common for other non-rendering processes to co-exist in the same computing environment. Processes from the operating system kernel may create minor disturbances of rendering because they share memory and CPU resources. Even though the modelling process for the rendering application does not account for such disturbances, the PID control action is expected to nullify them and continue to keep the frame rate stable.

If a large disturbance is introduced into the system through some unknown process, the rendering application may suffer a huge momentary fluctuation in its frame rate. In this case, the PID controller may not be able to correct the error and thus allow the rendering process to swing beyond the controllable operating range.

We constructed the PID control system for the rendering process in MATLAB as shown in Figure 5.5. The key components are the PID controller block, the plant block, and the interlinking communication channels. In a simulation environment, we assume that the data transfer has zero latency because the control system is executed by the same computer in the same memory space. However, when the plant model is swapped with the actual rendering process, this assumption may not be suitable because of network latency and communication overhead that can affect the performance of the control system.

The difference between the two simulation environments is that both the plant and the controller run on the same computer. The second environment has both functions reside in different computers connected via a local area network (LAN). The communication blocks in the control system use the transmission control protocol (TCP) to send packet data over the network from the source to the destination location between the controller and plant. This network communication protocol was selected because of the guaranteed delivery mechanism to ensure that data streams between the plant and controller will not be dropped for every rendered frame.

In addition to the mutual loading problem caused by the plant and controller processes, the Windows operating system poses the limitation of multitasking in the graphical user interface (GUI) environment. This limitation prevents one application window from receiving prioritised CPU time if it does not receive the focus

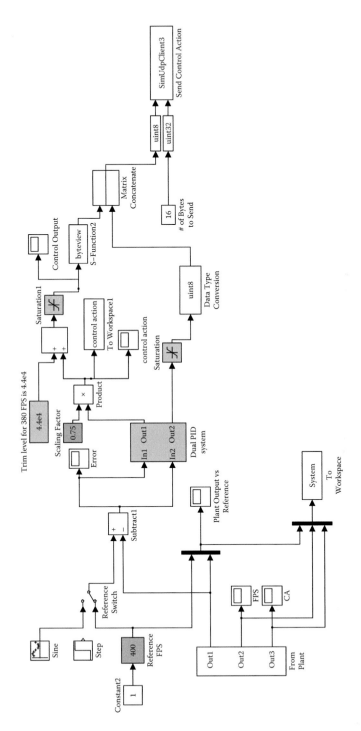

FIGURE 5.5   PID control system in MATLAB.

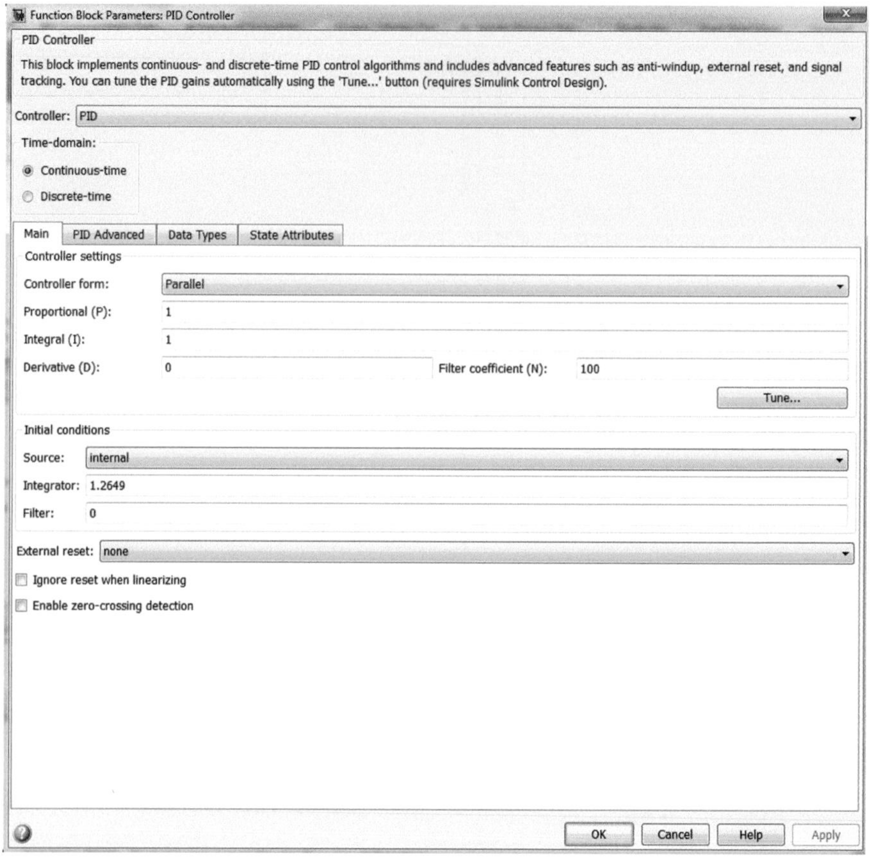

(a)

FIGURE 5.6 (See colour insert.) (a) Setting PID controller gain values in MATLAB. (b) Interactive graphical user interface in MATLAB/Simulink for tuning PID controller.

and restricts the plant and controller to execute in unison. This constraint made it imperative for our control system to be implemented across machines dedicated to the plant and controller processes separately.

To ensure that the PID controller is configured optimally, its gain values must be set correctly. Section 5.3 introduced the *tuning* process that is required before any PID controller can be employed. We tuned our PID controller using the Robust Response-Time Tuning Algorithm in MATLAB. Figures 5.6(a) and (b) illustrate the graphical user interface for this PID controller tuner.

The MATLAB/Simulink® PID controller block offers two key benefits for controller design. First, it integrates with the control design toolbox to provide closed-loop feedback analysis for linear systems. The resultant plots are very useful for analysis of various-system related considerations such as stability, frequency response, and step response. Second, along with the GUI shown in Figure 5.6, the

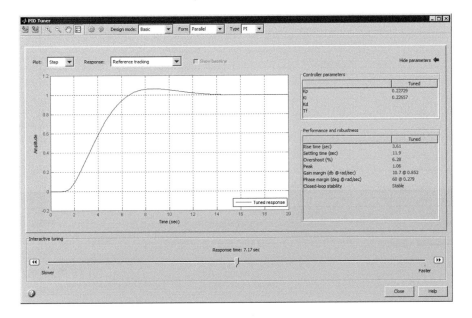

(b)

FIGURE 5.6 *(Continued)*

user has the flexibility to manually overwrite the PID parameters for trials to obtain better system performance. As a result, the PID controller design process is greatly accelerated.

5.3.2 DATA PREPROCESSING IN PID CONTROL SYSTEM

It is common that raw data values obtained from experiments can vary from very small fractions to extremely large numbers. For example, in real-time rendering, the geometry (vertices) required for the construction of a 3D object can yield numerical values ranging from a few thousand to hundreds of thousands. Computations involving very large or small numbers may be difficult because of the data format required to represent them precisely. Consequently modelling errors may occur and cause further errors in the controller design phase.

To avoid this problem, prescaling and normalisation of numerical values are often done. In a simulation environment, internal scaling can improve control system performance significantly. However, the trade-off is that the scaled values may not always be directly indicative of real world data values.

The scaling issues are relevant to data structures internal to the system model and the controller. In implementing a PID control system, an amplifier is usually adopted so that controller tuning process can be simplified using smaller numerical values. An amplifier is shown in Figure 5.5 as the multiplier block that resides at the output of the PID controller.

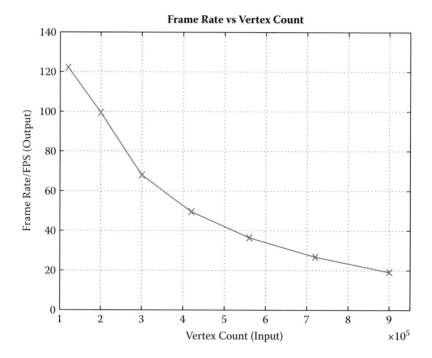

FIGURE 5.7   Steady-state frame time and vertex count relationship shown in Experiment 1.

5.3.3   Gain Scheduling for Non-Linear Rendering Process Models

As mentioned in Chapter 4, the real-time rendering process exhibits non-linearity characteristics. Prior research [38] demonstrates that the time taken to render a vertex differs with changing rendering loads. Figure 5.7 (originally Figure 3.7 from Chapter 3) is reproduced here and clearly shows this property.

We can observe from the figure that a single line segment approximation of the system's input–output relationship is inadequate. One approach to the challenge of designing a control strategy to counter this problem is to resolve the non-linearity at a piecewise level. In other words, we can approximate the system's input–output relationship with a series of line segments at selected operating points instead of using a single line across an entire operating range.

Each line segment represents a region whereby the plant may be modelled using linear model structures. Thereafter a suitable PID controller can be designed and introduced to achieve the desired performance for a delta region near a particular operating point. It is important to note the intention of Figure 5.7 is not to dictate or convey the number of segments to use for any particular application. It is produced to verify the existence of non-linearity in the rendering process. In practice, the number of linear segments to use is dependent and specific to the user's modelling requirements.

Because numerous combinations of line segments can approximate the curve shown in Figure 5.7, we can approach this optimal allocation of line segments to describe the non-linear input–output relationship as a constrained optimisation problem with a minimal number of line segments. First, we present this non-linear relationship represented by a polynomial model:

$$y = \sum_{i=1}^{n+1} p_i x^{n+1-i} \ , \ u_0 \le x \le u_N \tag{5.3}$$

where $(n + 1)$ is the order of the polynomial and n is the degree of the polynomial. The order denotes the number of coefficients to be fit, and the degree represents the highest power of the predictor variable. Since straight line segments are used to fit the curve, the degree of the polynomial is chosen as 1. The objective is to derive a series of line segments that fulfill the approximation of this relationship by:

$$y = \begin{cases} a_1 + b_1 x_1, & u_0 \le x \le u_1 \\ a_2 + b_2 x_2, & u_1 \le x \le u_2 \\ \quad \cdots & \quad \cdots \\ a_N + b_N x_N, & u_{N-1} \le x \le u_N \end{cases} \tag{5.4}$$

where the variables a and b minimise the following equation:

$$F\left(a_1, a_2, ..., a_N, b_1, b_2, ..., b_N, u_1, u_2, ..., u_{N-1}\right)$$

$$= \sum_{j=1}^{N} \int_{u_{j-1}}^{u_j} \left(f(x) - a_j - b_j x\right)^2 dx \tag{5.5}$$

The right side of the equation represents the least square error of the approximation. The different approaches to solving this problem are offered in previous research by Stone [58], Bellman [59], and Chan and Chin [60].

From the linear ranges derived, the corresponding input–output data set is used for model identification. The model structure is represented by the state–space [57] Equations [(5.6) and (5.7)]. The parameters of this system model structure may be obtained using the subspace algorithm (N4SID) [1].

$$x(k+1) = Ax(k) + Bu(k) \tag{5.6}$$

$$y(k) = Cx(k) + Du(k) \tag{5.7}$$

Based on the non-linear operation characteristics of the rendering process, a single PID controller would be inadequate to provide reasonable control performance over the entire operating range. Therefore we approach the problem by scheduling different gain values for the PID controller according to the respective linear operating ranges (Figure 5.8). In this design, the configuration of the PID controller can be stored in a look-up table so that the relevant values may be set into the PID controller

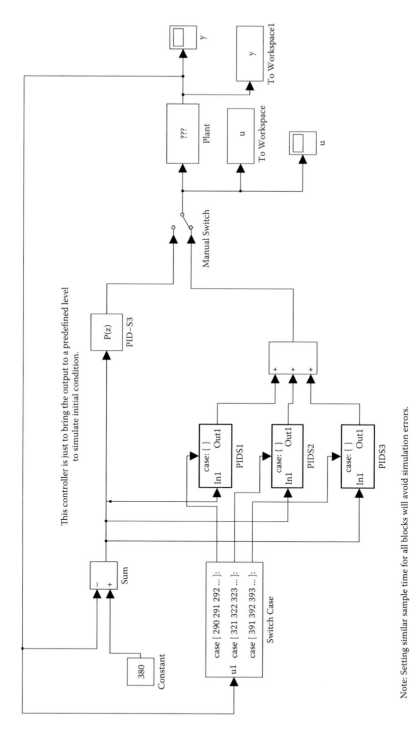

FIGURE 5.8   Gain scheduling PID control system.

as it traverses the different operating ranges. Such an implementation may be realised in MATLAB/Simulink as shown in Figure 5.8.

Since the PID gain values may differ greatly over the entire operating range, scheduling these values directly into the PID controller may cause some unexpected jitter at the border where the switch takes place. This jitter is typically manifested as a disruption to the control mechanism and may create some inconsistencies in plant output. A common technique to reduce the effect of this jitter is to adopt linear interpolation between the controller's gain values.

5.3.4 NEURAL PID CONTROL

The PID controller is computationally straightforward and effective. The challenge lies in tuning its gain parameters, especially when it is difficult to derive a system model. In the previous section, we examined the use of a combination of separately tuned PID controllers. The goal is to create a control system that works over a large operating range and is resilient to the effects of the system's non-linear characteristics.

In this section, we investigate a technique that does not require the cascading of PID controllers and eliminates the effort to tune them separately. This technique also allows a single PID controller to be continuously tuned online while the system operates.

Artificial neural networks (ANNs) are well known to be capable of memory retention and learning through their adaptive nature of modelling non-linear functions. By utilising an artificial neuron to learn and adaptively tune a PID controller in the single neuron adaptive PID (SNPID) control algorithm [64], it is possible to achieve continuous control with good performance over a substantially large operating range. Figure 5.9 illustrates the SNPID control system design. Recall from Section 5.3 that the discrete incremental PID controller may be expressed as:

$$u(k) = K_p e(k) + K_i \sum_{k=0}^{n} e(k) + K_d (e(k) - e(k-1)) \tag{5.8}$$

where $u(k)$ is the output of the controller and K_p, K_d, and K_i are the proportional gain, derivative gain, and integration gain, respectively. $e(k)$ is the error between the reference and system outputs that serves as input to the controller. In the SNPID control implementation [64], we have the output of the neuron given by the following equation.

$$Y = X^T W \tag{5.9}$$

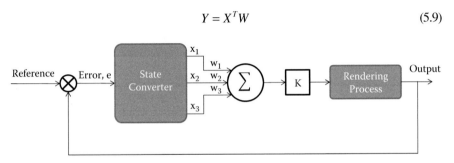

FIGURE 5.9 Single neuron PID control system.

where Y is the output of the neuron, X represents the internal states of the neuron, and W denotes weights assigned to each neural connection given as $X = (x_1, x_2, x_3)$ and $W = (w_1, w_2, w_3)^T$, respectively. In the SNPID controller configuration, the adaptive weights w_i are analogous to the conventional PID gains, K_p, K_i, and K_d. From Equation (5.8), the output of the SNPID controller is further expressed as

$$u(k) = K \sum_{i=0}^{3} w_i(k) x_i(k) \tag{5.10}$$

where K is the gain value of the neuron. Note that the input to the neuron at time k is given by the following equations.

$$x_1(k) = e(k) - e(k-1) \tag{5.11}$$

$$x_2(k) = e(k) \tag{5.12}$$

$$x_3(k) = e(k) - 2e(k-1) + e(k-2) \tag{5.13}$$

The errors at time k, $(k-1)$, $(k-2)$, etc., are represented by $e(k)$, $e(k-1)$, $(k-2)$, etc. where:

$$e(k) = r(k) - y(k) \tag{5.14}$$

Hence the single neuron PID control law may be expressed as:

$$u(k) = u(k-1) + K \sum_{i=0}^{3} w_i'(k) x_i(k) \tag{5.15}$$

whereby the weights are determined by the Hebb learning algorithm described below from Equations (5.16) to (5.19).

$$w_i'(k) = \frac{w_i(k)}{\sum_{i=1}^{3} |w_i(k)|} \tag{5.16}$$

$$w_1(k) = w_1(k-1) + \rho_1 e(k) u(k-1)(2e(k) - e(k-1)) \tag{5.17}$$

$$w_2(k) = w_2(k-1) + \rho_2 e(k) u(k-1)(2e(k) - e(k-1)) \tag{5.18}$$

$$w_3(k) = w_3(k-1) + \rho_3 e(k) u(k-1)(2e(k) - e(k-1)) \tag{5.19}$$

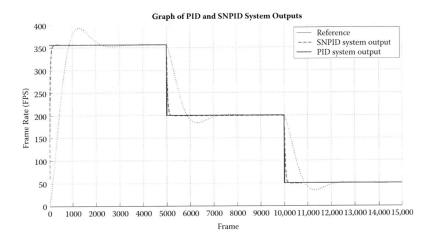

FIGURE 5.10   Comparison of system outputs using SNPID and PID controllers.

Experiment—To compare the performances of the SNPID and PID controllers, we created a MATLAB simulation consisting of a closed-loop feedback system using the system model derived in Experiment 3 from Chapter 3, Section 3.5.3. The system first ran with the SNPID controller, then with the PID controller over the same duration of 15,000 frames. During the simulation, we allowed the system output to stabilise before a new reference was set. In contrast to previous modelling experiments, we wanted to validate the performance of the SNPID controller over a large operating range. To achieve this, we deliberately set the reference changes in bigger steps.

Simulation results—Figure 5.10 presents the simulated system outputs from the SNPID and PID controllers. Figure 5.11 shows their respective and corresponding control actions. Figure 5.10 indicates that the SNPID controller provides faster response than the PID controller with almost no overshoot at the system output. The SNPID controller was approximately two times faster than the PID controller in reaching new steady-state references.

5.4   EXPERIMENTS

In this section, we present the details of the experiments conducted to validate the control framework described in this chapter. The objective was to demonstrate the implementation of a closed-loop feedback control system involving the real-time rendering process with the plant and PID controller. The rendering process to be controlled was the same as the application mentioned in Chapter 3. Figure 5.12 is a screenshot of the application. The details on deriving the system model and its parameters are also provided in Chapter 3.

Using the derived linear model, we first described two experiments in MATLAB: one executed fully in a simulation environment and the other an actual rendering process. In the actual rendering experiment, the plant and controller run on separate computers as described in Section 5.2.1. Two sets of data were collected. Each

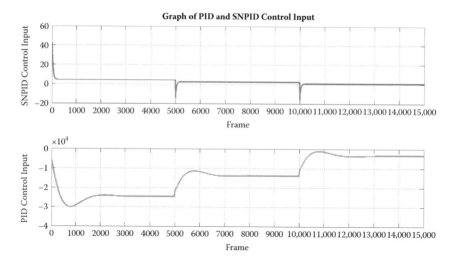

FIGURE 5.11   Control input from SNPID and PID controller.

captured the same input (vertex count) and output (frame rate) quantities of the rendering process. The third experiment dealt with the application of the gain scheduling control system described in Section 5.3.3 in a simulated environment.

All experiments were run on a desktop computer with an Intel Core 2 Quad CPU at 3 GHz, with 8 GB of main memory and NVIDIA GeForce GT 320 graphics processor hardware (with 4 GB video memory) on a 64-bit Windows 7 operating system. The system identification toolbox was used for deriving the linear models of the rendering processes. The control design toolbox was used for the design and analysis of the feedback control systems.

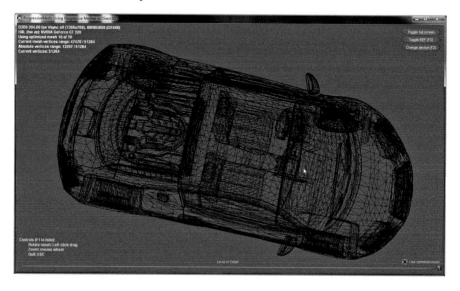

FIGURE 5.12   **(See colour insert.)** Screenshot of application with PID control.

5.5 RESULTS

5.5.1 SIMULATION ENVIRONMENT

The results of the first experiment are shown in Figures 5.13 and 5.14. Based on Figure 5.13, the objective was to allow the controller to track a higher reference level. The rendering process was allowed to stabilise at approximately 360 FPS before triggering of a new reference level of 410. The PID controller took approximately 2,500 frames (6 s) to reach the target reference level. Furthermore, almost no overshooting was observed. A small amount (fewer than 5 FPS) of tracking error was noted. At a high frame rate of 400 FPS, this error can be regarded as negligible.

The PID control system was then tested for its ability to track a lower reference level (380 FPS) from an initially higher frame rate (400 FPS). The results are shown in Figure 5.14. Tracking was very accurate (error rate below 3 FPS) and no oscillation arose from the control action.

5.5.2 CONTROL SYSTEM WITH ACTUAL RENDERING PROCESS

The results of the second experiment in which the plant model is replaced by the actual rendering process are presented in Figures 5.15 and 5.16.

In the first part of the second experiment, the rendering process was allowed to stabilise at 350 FPS before a trigger changed the reference level to 390 FPS. The PID control action took approximately 25,000 frames to reach a stable frame rate close to the reference level. The steady-state error was approximately 5 FPS. In the second part of the same experiment, the control system was allowed to track a 380 FPS

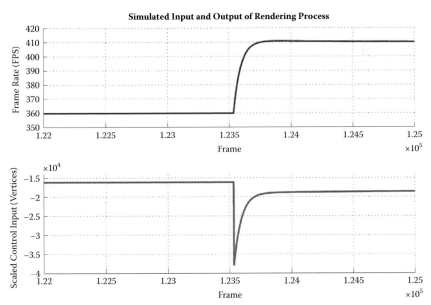

FIGURE 5.13 Reference tracking using PID controller (low to high).

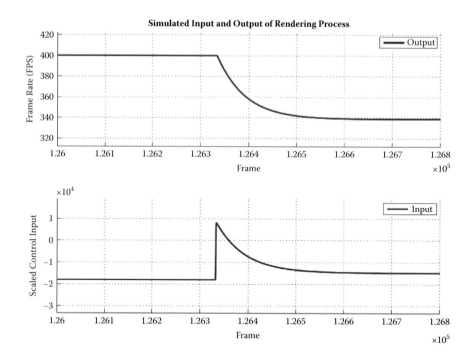

FIGURE 5.14 Reference tracking using PID controller (high to low).

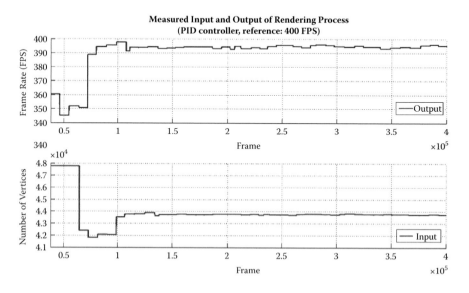

FIGURE 5.15 Reference tracking using PID controller (to higher FPS).

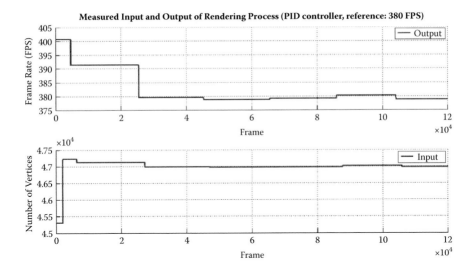

FIGURE 5.16   Reference tracking using PID controller (to lower FPS).

reference level lower than its initial reference rate (400 FPS). We can see that the controller takes approximately the same time (25,000 frames) to bring the frame rate down to the lower reference level.

Both phases of the second experiment showed consistency in plant response time and the control action was able to track the reference level eventually. This indicates that the system model representation is reasonably accurate and that the PID controller was tuned adequately to work with the actual rendering process.

5.5.3   GAIN SCHEDULING CONTROL SYSTEM

With reference to Section 5.3.3 covering the application of PID control over an operating range involving non-linear characteristics, we wanted to validate the applicability of the gain scheduling control strategy in such a scenario. First, an extended operating range was selected and divided into three segments as shown in Table 5.1. We modelled the rendering process within these segments at various operating points. A corresponding controller was designed at every operating range and its parameters were preset into the PID controller shown in Figure 5.8 for the purpose

TABLE 5.1
Linear Operating Ranges

Linear Range Reference	Approximate Frame Rate Range
1	390–450
2	389–325
3	324–280

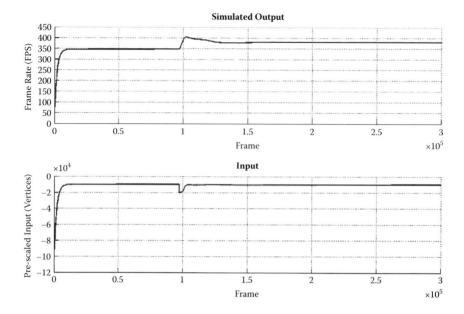

FIGURE 5.17   Simulated output with gain scheduling PID controller.

of switching the PID gain values as the controller processed output errors across the extended operating range.

The first objective of the rendering system was tracking a predefined level at 340 FPS. This is an output level in Operation Range 2 from Table 5.1. Figure 5.14 indicates that the controller can bring the output to this level quickly with no over-shoot and track the level steadily. Thereafter, we wanted to observe the controller's ability to track a new reference level in a different operation range. We set the new reference output level to 390 FPS (Operation Range 1 in Table 5.1). The new reference output level was set while the control system was running.

Figure 5.17 shows the tracking of a new reference level with some overshoot. To drive the output toward the target reference level, the control action caused the input to the plant to take a steep dip that indicates an abrupt over-correction. Nevertheless, the rendering process output was still brought to the desired reference level with negligible error (fewer than 2 FPS).

5.6   CONCLUSION

In this chapter, we introduced the concept of using control principles to track real-time rendering performance. The controller design was based on a closed-loop feedback system with a plant model. Although no restrictions were imposed on the controller design, we utilised the PID algorithm as the control strategy in a real-time rendering application.

Since real-time rendering is inherently non-linear, we provided a solution to control this process from a piecewise approach by approximating a large operating range by grouping smaller linear ones. We also introduced the neural-assisted

PID control technique which is a superior approach to the conventional PID control design. It does not require manual tuning of its gain parameters and constitutes a viable solution for achieving performance targets in real-time rendering

In summary, we have shown by our experiments that the PID controller is effective in keeping the output of a rendering process close to the user defined performance target for linear rendering system models and when gain scheduling is adopted for larger operating ranges.

6 Model-Less Control

6.1 INTRODUCTION

In this chapter we consider a different perspective for controlling the rendering process. While conventional data-driven system identification strategies may be adopted to derive a rendering process model, the result may not necessarily imply that an accurate model can be derived without resolving certain technical challenges in data processing.

To circumvent such problems, this chapter investigates an approach to controlling the rendering process by allowing the user to exploit *a priori* information about the rendering process without the need for an explicit rendering model by using a soft computing method known as fuzzy control. The fundamentals of fuzzy set theory and the mathematics for a conventional fuzzy inference system are provided in Section 4.2.2 in Chapter 4.

6.2 FUZZY CONTROL SYSTEM

The construction of a fuzzy logic control system is relatively similar to the PID control system described in Chapter 5. Based on the same architecture described by Figure 5.2 in that chapter, we introduce the fuzzy controller is used in place of the PID controller. As in the case of the PID controller in which the quantity of the input to the plant is varied directly, no strict rule applies to the selection of the input to a fuzzy control system. Certain fuzzy control systems such as applications for temperature and process control utilise the rate of change of the input to the plant instead of the numerical value of the quantity. In this research, the rate and the direction of change (increase or decrease of vertex count) are used.

The design of a fuzzy control system consists of two phases. First, we develop the fuzzy control system in a simulation environment where the plant model is used. After the control system is validated to work correctly, we replace the plant model with the actual rendering process as done in previous experiments.

Unlike a PID control system, a fuzzy logic controller functions on linguistics variables instead of numerical values. As described in Section 4.2.2, a fuzzy logic system is defined primarily by the type or structure of the controller, the rule base, and the membership functions of the input and output of the process to be controlled (Figure 6.1). In this research, we adopted the Mamdani fuzzy model. The rule base was constructed based on a straightforward inverse input–output relationship between the frame rate and the rate of change of vertex count. This fuzzy inference rule set is shown in Table 6.1.

The fuzzy logic toolbox in Simulink®/MATLAB® provides comprehensive tools such as the rule editor and membership function editor to accelerate the

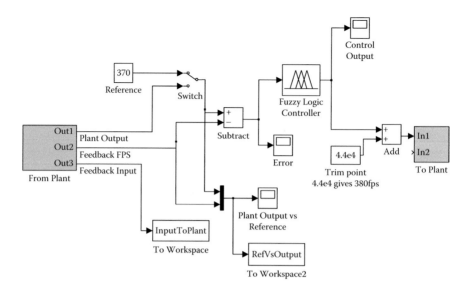

FIGURE 6.1   Fuzzy control system in Simulink/MATLAB.

TABLE 6.1
Fuzzy Inference Rule Set

If	Then
fps_error IS High	vertex_count IS DecreaseHigh
fps_error IS Low	vertex_count IS IncreaseHigh

implementation of fuzzy control. Figure 6.2 illustrates the graphical user interface that allows the creation of membership functions, the set-up of the rule base, and several other parameters that may be changed.

Fuzzy logic deals with non-crisp values. Thus the approach to tuning fuzzy logic controller parameters relies on heuristics and iterative processes that allow easy observation of the effects on simulation performance from changes in fuzzy controller parameters. Some parameters that may be changed include the membership function and the membership input and output ranges.

The fuzzy logic toolbox provides a step-through functionality in simulation time. This allows a user to observe how a defuzzified decision is derived by viewing the fuzzified inputs and how they are combined to produce the output via the firing function. This tool is important for helping a user decide the appropriate membership function to use by analysing the output of the fuzzy controller over a series of steps.

6.3   ADAPTIVE NEURAL FUZZY CONTROL

We described the basic structure of the type of fuzzy inference system in a systematic manner. In brief, it consists of multi-tier relationships that first map input

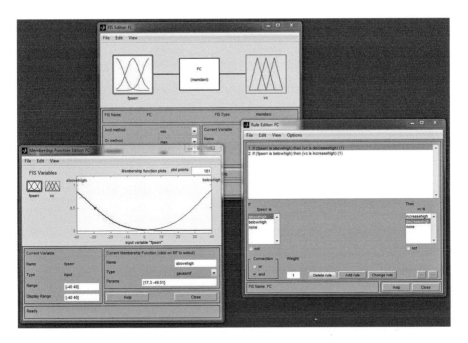

FIGURE 6.2   (See colour insert.) Configuring fuzzy controller in Simulink/MATLAB.

characteristics to their membership functions. Thereafter, each input membership function is mapped to rules that are correlated to a set of output characteristics. Finally the output is determined by traversing the relationship between the output characteristics and the output membership functions such that a crisp or single-valued output is produced.

One key element in a fuzzy control system is the use of fixed membership functions that were chosen arbitrarily. In other words, the applied fuzzy inference is applied only to systems whose rule structures are essentially predetermined by the user's understanding and interpretation of the characteristics of the variables in the system model.

Nevertheless, it is possible that collection of input and output data is available for modelling but it is not clear to the user whether a predetermined model structure may be appropriate based on the characteristics of the variables in the system. In certain modelling situations, it may not be possible to discern the correct membership functions to adopt by simply observing input and output data. Based on these scenarios, we approached the model-less control problem by using the adaptive neuro-fuzzy inference technique.

In brief, the neuro-adaptive learning technique that performs similarly to neural networks provides a method for a fuzzy modelling procedure to learn information from an input–output data set. By using the fuzzy logic toolbox in Simulink/MATLAB, it is possible to compute the membership function parameters that best allow the associated fuzzy inference system to track the given input–output data.

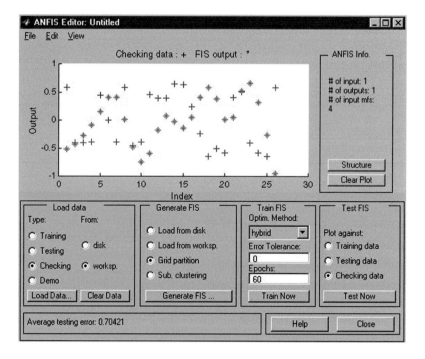

FIGURE 6.3   (See colour insert.) ANFIS editor graphical user interface in Simulink/MATLAB.

In contrast to the fuzzy controller design process in which the user selects the membership functions, the adaptive neuro-fuzzy inference system (ANFIS) is capable of constructing a fuzzy system whose membership function parameters are tuned via a back-propagation algorithm alone or in combination with a least squares method by using a specific input–output dataset. In other words, the resultant ANFIS embodies the modelling of the plant process through the construction of membership functions and their inherent relationships. Figures 6.3 and 6.4 show the graphical user interface of the ANFIS tool in Simulink/MATLAB.

The process of constructing an ANFIS control system is similar to constructing a fuzzy controller except that the plant model is not used explicitly. From the same input–output data set collected, a portion is allocated for training while the rest is used for validation. After the ANFIS parameters such as the numbers of inputs, outputs, and membership functions are set by the user, the software computes the ANFIS structure that may be imported into a simulation environment for testing with the actual process. Figure 6.5 illustrates the control system design.

6.4   EXPERIMENT

We designed two experiments to validate our control system framework using a fuzzy controller. As in the previous chapter's experiments, our first experiment was performed in a fully synthetic simulation environment. The second experiment involved switching the plant with the actual rendering process.

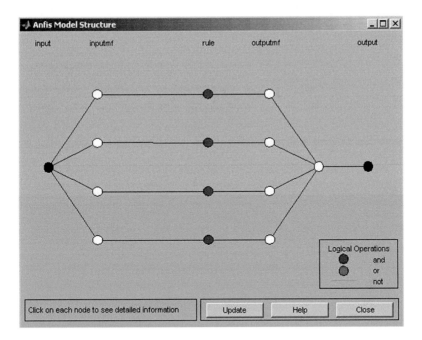

FIGURE 6.4   (See colour insert.) Neural network model structure in ANFIS.

The fuzzy rule-set used in these experiments is shown in Table 6.1. The selection of an appropriate membership function is non-automatic and we begin with a generic non-linear segment of a parabolic curve. The next step is testing the fuzzy controller with the actual rendering process over several iterations to ensure that both the gradient of the selected range and membership enrollment (curve) functions are suitable to perform the tracking function correctly. The membership functions are shown in Figure 6.6.

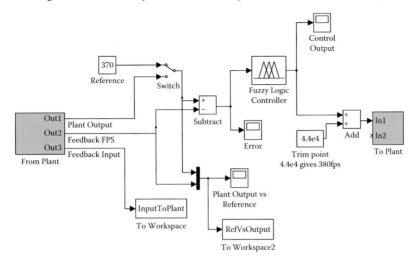

FIGURE 6.5   Using ANFIS for controlling real-time rendering process.

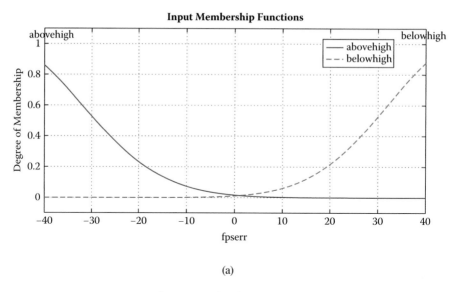

(a)

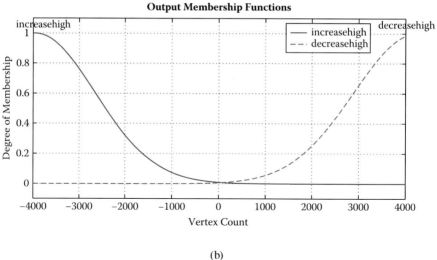

(b)

FIGURE 6.6 Input and output membership functions.

The *fps_error* fuzzy variable is set at ±40 FPS from the operating point of 400 FPS while the *vertex_count* variable is set at ±4,000 vertices. The fuzzy control system follows the design in Figure 6.7.

All experiments were run on a desktop computer with an Intel Core 2 Quad CPU at 3 GHz, with 8 GB of main memory and NVIDIA GeForce GT 320 graphics processor hardware (4 GB video memory) on a 64-bit Windows 7 operating system. The fuzzy logic toolbox was used to generate the fuzzy inference system and fuzzy controller design.

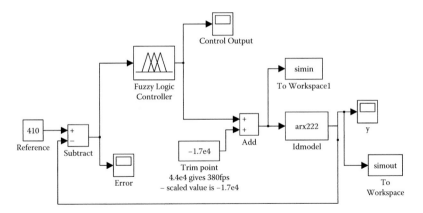

FIGURE 6.7   Fuzzy logic control system.

6.5   RESULTS

6.5.1   SIMULATION

Figures 6.8 and 6.9 show results from both parts of the experiment using a fuzzy controller to track frame rate level changes. In both scenarios, the controller was capable of varying the input to the plant so that its output followed the reference level closely. However, the fuzzy controller took longer to perform this task as indicated by the output response times in both parts of this experiment.

6.5.2   FUZZY CONTROL SYSTEM WITH RENDERING PROCESS

In the second experiment, the rendering process was allowed to stabilise at 400 FPS before the new reference level of 370 FPS was set as shown in Figure 6.10. The fuzzy

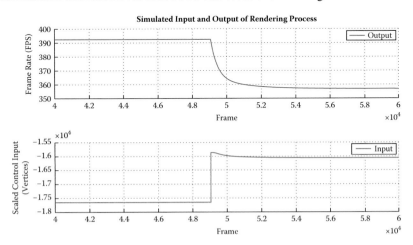

FIGURE 6.8   Reference tracking using fuzzy controller (high to low).

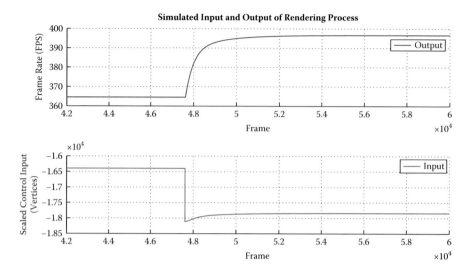

FIGURE 6.9   Reference tracking using fuzzy controller (low to high).

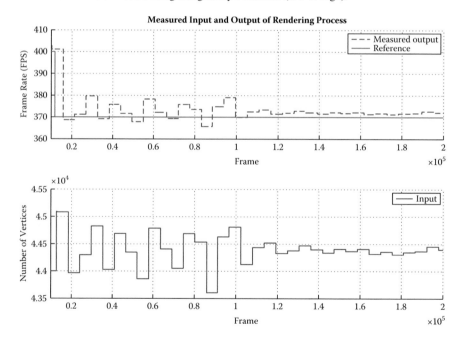

FIGURE 6.10   Reference tracking using fuzzy controller (to lower FPS).

logic controller was able to reduce the frame rate after adjusting the load of the rendering process. This adjustment took place over approximately 9,000 frames with an observable error of about 3 FPS.

In Figure 6.11, the frame rate increases from 370 FPS to a target of 400 FPS. Again, the fuzzy logic controller can reduce the load and track the new reference

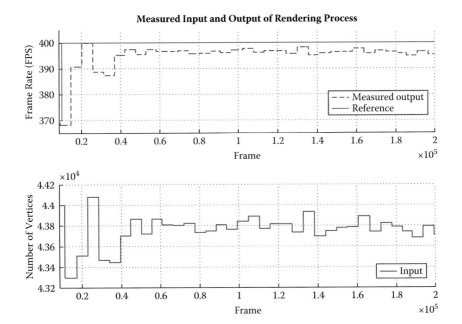

FIGURE 6.11   Reference tracking using fuzzy controller (to higher FPS).

frame rate with an approximate error of 5 FPS. The response for this action takes approximately 4,000 frames—fewer than required by the previous control action. However, in terms of tracking accuracy, the error can be observed as slightly larger than the one shown in Figure 6.10. In both figures, some fluctuations of input to the rendering process are observed. This can be explained by the scaling computation used in the control framework and the resolution of the computer program data structures.

In addition to comparing differences in controller designs, another objective of using the ANFIS controller was to determine its robustness for handling variations in user-defined references. Figure 6.12 depicts examples of such variations. Note that the variations may span over an operating point or zone where the rendering process may be approximated by a linear model. The ability of the ANFIS controller to maintain the plant output close to the changing reference levels in such a scenario indicates that it is inherently capable of controlling non-linear rendering processes.

6.6   DISCUSSION

Figure 6.11 indicates that the rendering process output tracks the user set reference only after a short delay. This can be explained by the experiment set-up involving network communication. Since the plant and controller communicate via a network connection, transport delays arising when data are sent between the plant and the controller are expected.

While such delays are minimised via an isolated network infrastructure, it should be noted that data transport within computer systems is not instantaneous. Such

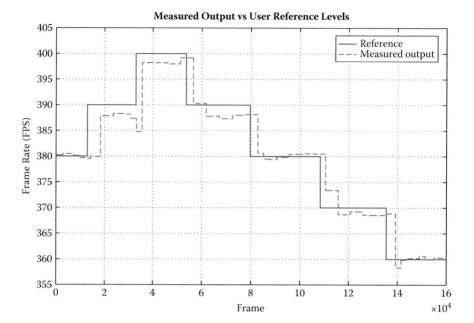

FIGURE 6.12 Continuous reference tracking using ANFIS controller.

delays could be caused by different CPU utilisation and wait states or transports between memory storage locations.

The impact of delay on a control system should not affect a system's ability to track user-defined performance targets noticeably. If a delay does affect the overall effectiveness of a control system, its effect should be modelled as part of the system as well. A detailed investigation into the intricacies of delay is beyond the scope of this book. Our experiment yielded satisfactory results and further work on more elaborate distributed control systems may further extend this research.

6.7   CONCLUSION

The focus of this chapter is controlling real-time rendering via a model-less approach defined as no need for devising a formal mathematical representation of a rendering process plant model by which a rigorous system identification procedure is to be carried out. We provided a framework for model-less control of real-time rendering using a conventional fuzzy controller and adaptive neuro-fuzzy inference system.

The experiments showed that both techniques are capable of yielding good results without plant models. More importantly, we demonstrated that a model-less control framework can be extended to support a wider operating range where non-linear characteristics of the rendering process may appear.

7 Applications, Challenges, and Possibilities

In this chapter, we examine further the details of the implementation of control technology for computer graphics introduced in this book. The objective is to provide insights into practical aspects of designing such systems that we believe will allow this book to serve as a useful resource to practitioners in the control engineering and computer graphics fields.

7.1 SYSTEM ARCHITECTURES

The plant and controller system architecture described throughout this book may be realised in several forms, depending on the application and performance requirements. We classify the forms in broad terms into three categories:

A. Plant and controller in the same computer in the same process (different execution threads)
B. Plant and controller in the same computer and in separate processes
C. Plant and controller in separate computers

As a quick primer, a process in computer programming terminology is the execution of an instance of an application. A thread is a single path of execution within a process. In addition, a process (essentially an application) can spawn and use multiple threads. Since a process can consist of multiple threads, a thread is commonly classified as a lightweight process. Often, the essential differentiating point between a thread and a process is the nature of the task assigned to be accomplished.

Traditionally, developers use threads for smaller and specialised tasks such as network communication and to achieve parallelism in application design. In contrast, processes are used for heavyweight tasks and involve broader scopes encompassing most other subtasks of an application. Another important fact is that threads within the same process share the same memory address space, whereas processes do not. This implies faster execution for threads because it allows them to read from and write to the same data structures, facilitating speedier communications between them.

In Configuration A, the controller is built into the rendering application. With this architecture, the implementation of the controller's design must follow the programming language by which the rendering application is developed. In other words, the developer must use the same programming language as the rendering application to code the controller. Since the plant and controller are compiled and built into the same binary, this configuration provides fast speed for runtime execution.

However, this configuration also demands separation of the plant and controller into different execution threads because the rendering process must run asynchronously and in tandem with the controller thread. A common mistake in structuring the code is to have the controller implemented in the same process. The results may be significantly slower frame rates due to hogging of CPU time by the controller when it shares a CPU time slice with the plant.

In Configuration B, the plant and controller execute as separate process within the same machine. A simple illustration is the use of two applications to realise this architecture, for example, using a specialised controller application that can communicate with the rendering application in real time. This is usually done via inter-process or network communications within the local computer.

Since rendering processes vary in complexity and computational requirements, Configuration C provides a system architecture that decouples the plant and controller. This means that the controller can execute properly even if the plant computer is not suitable for running run both processes. Furthermore, Configuration C allows the control system to run without subjecting each process to the limitations of the underlying operating system (e.g., application window handling) and helps minimise the effects of kernel process disturbances on the control system. We conducted the experiments described in this chapter using Configuration C because:

- The controller was designed and simulated in Simulink®/MATLAB® and run as part of the main Simulink/MATLAB application.
- Due to development practices, MATLAB provided several useful graphing windows for debugging and performance monitoring. Configuration B could not be adopted because the Windows operating system does not support multiple visible windows with equal CPU usage priority—at least not with MATLAB application windows and our rendering application. Using Configuration B for our control system implementation would have kept the controller or plant from running properly.
- By using separate and dedicated computers for the controller and plant, we circumvented the above problems.

A noteworthy point at this juncture is that our conduct of experiments using Configuration C does not suggest that the other two configurations are not useful. The adoption of any configuration depends on the system designer's choice of the optimal way to devise a control system based on application requirements, nature of system, components to be used, and time and effort available.

Figure 7.1 is a timing diagram of our plant (rendering) application. Note that a thread is created for network communication at the beginning of the application and kept running until the end of the application's life. The intent is to prevent the fetching and sending of data from and to a remote computer from bogging down the main rendering process. After the application initialises all required resources, the network communication thread works with the main process rendering loop at every frame to ensure that the control action is parsed and put into effect as needed. In addition, the controller also receives feedback information—essentially the frame rate from the plant for processing.

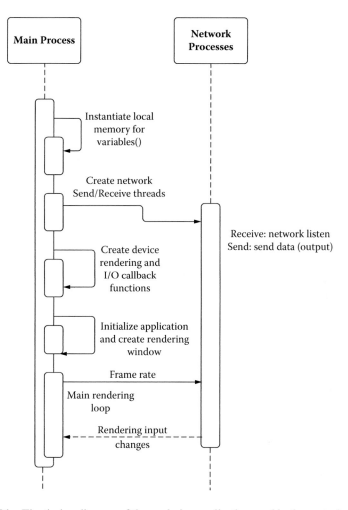

FIGURE 7.1   The timing diagram of the rendering application used in the control system.

7.1.1   Software Design

One key requirement for implementing a control system involving a real-time rendering application is that code level access is a primary task for the system to achieve the following:

1. A form of interprocess or shared memory communication can be established between the rendering application and the controller.
2. Input and output data points from selected variables can be tapped.
3. A form of rendering load control can be implemented.

Figure 7.2 shows a high-level componentised view of the rendering application used in the experiments in this chapter. In addition to the main rendering components, other

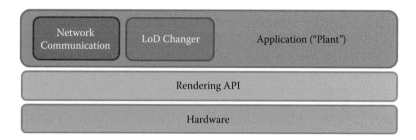

FIGURE 7.2   **(See colour insert.)** The high-level design of the rendering application.

essential software modules include the network communication and level-of-detail (LoD) control blocks. LoD blocks are just example and may well be replaced with other blocks capable of varying the rendering load to alter system output.

The layered structure of the design in Figure 7.2 illustrates a service-oriented approach; each layer makes use of the service offered by the layer below it. While the network communication and LoD changer blocks interact with the main application (real-time rendering process), the main application makes use of the rendering API residing in the operating system. Furthermore, the rendering API then utilises the computer hardware to perform the final rendering functions that lead to the generation of visible pixels on a display device.

To exemplify the application design shown in Figures 7.1 and 7.2, we draw a direct reference to this architecture with the C++ code of the application used in Experiment 3 in Section 3.6.3 (see also Annex A). The following correspondences to the architecture are highlighted:

1. Two threads instead of one, each dedicated to sending data and receiving data from the network as shown by *SendDataThreadFunction* and *ReceiveDataThreadFunction*.
2. The LoD changer block is synonymous with *SetTriangleCount, SetShaderComplexity*, and *SetNumVertices* functions.
3. The application makes use of the DirectX-rendering API to provide hardware-accelerated rendering in real time.

Possible architectural abstractions—So far, we described implementation designs for the controller at the application level, that is, the controller is in either process or thread form. Despite this, it is certainly possible to have different abstractions for the implementation of the controller:

1. The controller may be implemented as an API for users to integrate directly into their existing real-time rendering application.
2. The controller may be integrated into the LoD function in rendering applications. The appeal of this implementation is that the user has no need to handle programming interfaces; the control mechanism is fully automated. One possibility is to have the controller deeply embedded in short programs that are loaded into the graphics processor during runtime [1].

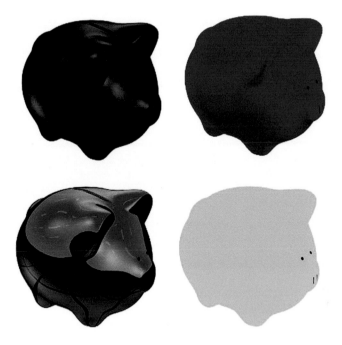

FIGURE 2.4   Samples of surface shading effects that can be achieved with pixel programs.

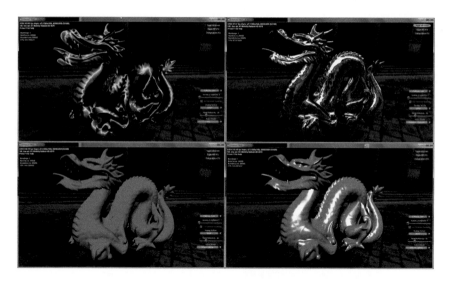

FIGURE 3.3 Screenshot of hardware tessellation sample application from DirectX SDK adapted with Stanford Dragon model in Experiments 1 and 2.

FIGURE 3.4 Screenshot of application in Experiment 1.

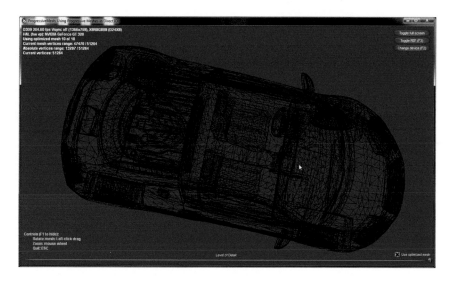

FIGURE 3.5 Screenshot of application in Experiment 3.

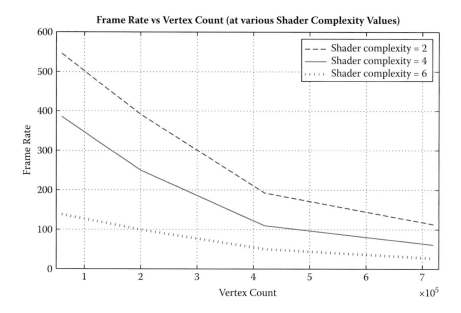

FIGURE 3.10 Steady-state outputs of the system based on selected combinations of two input variables.

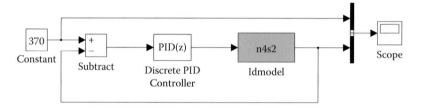

FIGURE 3.15 SISO control system in Experiment 3.

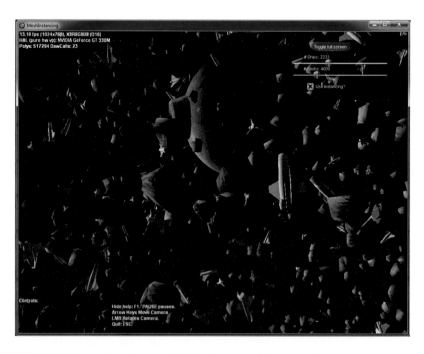

FIGURE 3.18 Screenshot of test application in superposition experiment.

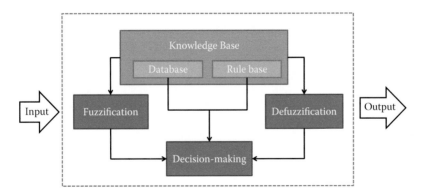

FIGURE 4.3   Fuzzy inference system.

FIGURE 4.4   Screenshot of application in Experiment 1.

FIGURE 4.5 Screenshot of application in Experiment 2.

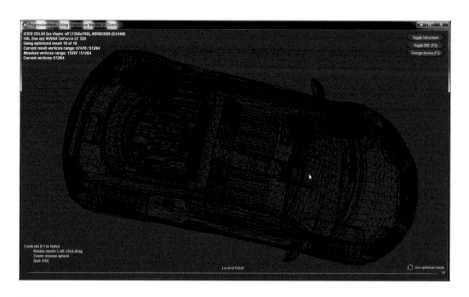

FIGURE 4.10 Screenshot of rendering application in Experiment 3.

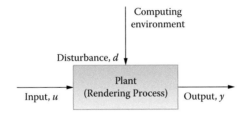

FIGURE 5.1 Rendering process from system perspective.

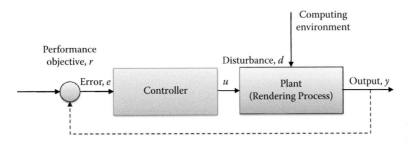

FIGURE 5.2 Closed-loop feedback control system.

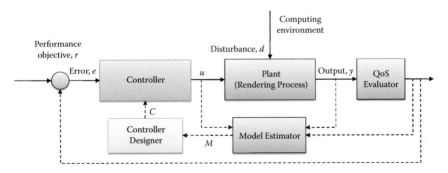

FIGURE 5.3 Rendering system with adaptive controller and quality of service feedback.

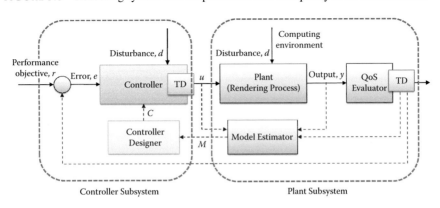

FIGURE 5.4 Modular adaptive control system for real-time rendering.

(a)

FIGURE 5.6 (a) Setting PID controller gain values in MATLAB. (b) Interactive graphical user interface in MATLAB/Simulink for tuning PID controller.

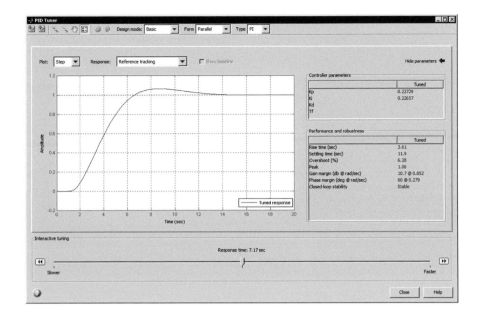

(b)

FIGURE 5.6 *(Continued)*

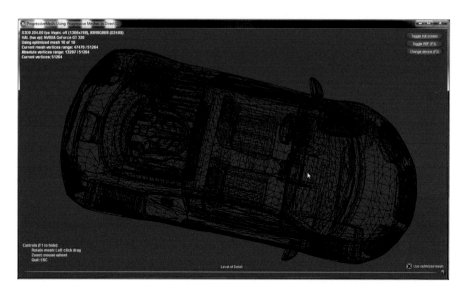

FIGURE 5.12 Screenshot of application with PID control.

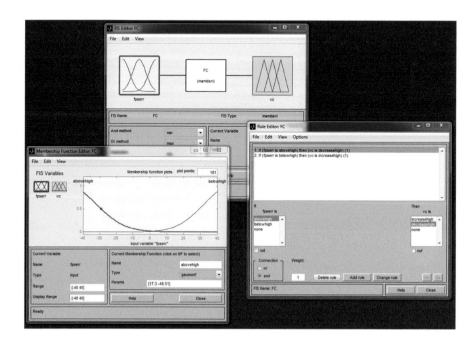

FIGURE 6.2 Configuring fuzzy controller in Simulink/MATLAB.

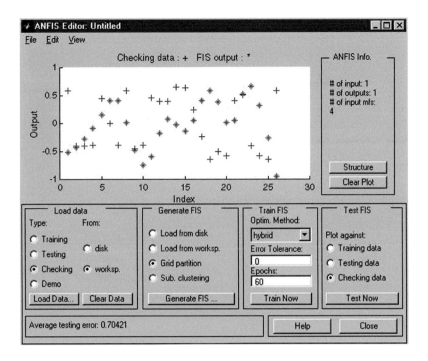

FIGURE 6.3 ANFIS editor graphical user interface in Simulink/MATLAB.

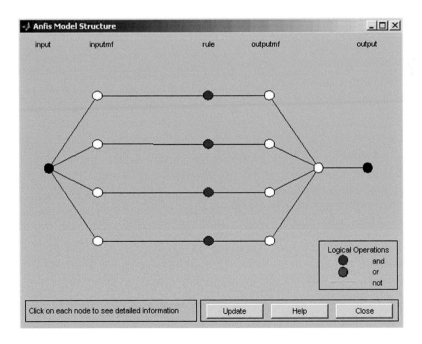

FIGURE 6.4 Neural network model structure in ANFIS.

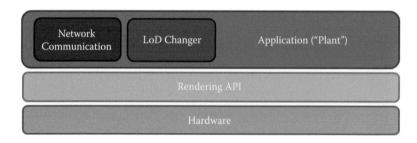

FIGURE 7.2 The high-level design of the rendering application.

7.2   SOFTWARE AND HARDWARE
PERFORMANCE CONSIDERATIONS

The process of establishing a valid control system for a real-time rendering application requires special attention beginning from data collection and pre-processing and proceeding to system level moderation. We provide a list of the key hardware and software considerations below.

7.2.1   DATA INTEGRITY

When a non-real-time operating system is used, it is inevitable for kernel processes to introduce disturbance to the rendering system output. Consequently, the data collected for system identification may contain spikes and occasional data points that do not follow the trend of the change in data direction. If such data are used for system identification directly, the result would be greater difficulty in deriving an accurate system model. Therefore, it is imperative to use a number of de-noising, filtering, and de-trending tools to preprocess data before the system identification step.

7.2.2   PLANT–CONTROLLER COMMUNICATION LATENCY

When network communication is involved as it is in Configurations B and C (described in Section 7.1), it is important to consider the network latency that may arise from closed-loop feedback data movement. For Configuration B, even if the interprocess communication is handled via a loop-back network communication, the latency would be negligible because of the local network hardware. In contrast, this assumption deviates more with Configuration C because of real data routing latency across network switches and other physical media. To minimise latency, the set-up should include a high-speed switch and local area isolated network. This arrangement was used in all our experiments.

From a software implementation perspective, the selected communication protocol plays an important role in system performance. In typical network communication arrangements the transmission control protocol (TCP) or universal datagram protocol (UDP) may be used. Configuration B would demonstrate a negligible difference between these options since performance is largely driven by hardware.

However, this is not true for systems using Configuration C. If the network connecting the plant and controller cannot be isolated for certain reasons, it would be better to use TCP as the mode of communication because it guarantees lossless delivery. Conversely, if a network is unlikely to be congested, UDP can be a good choice because it provides better speed.

7.2.3   DATA STRUCTURES AND HANDLING

For optimised transmission efficiency, it is always advantageous to use simple data structures. Complex data structures should be avoided to prevent data marshalling issues that may arise due to incompatibility in machine-specific hardware. Also, it is a good practice to utilise a single network communication channel for sending and one for receiving data as much as possible so that CPU utilisation and contextual switch overhead are optimised.

For performance reasons, data should be sent only when a change occurs. This latch-on technique allows both the plant and controller to run more efficiently without the need to waste CPU processing time or face network latency as long as the last sent or received value is valid.

Finally, data trim points are critical and necessary because they prevent spikes in data value due to conversion or other errors from destabilising the control system. If they are not implemented, the rendering process may produce unexpected outcomes such as substantial fluctuations in frame rates due to erroneous computation by the controller.

7.2.4 COMPLEXITY OF CONTROL ALGORITHM

As described in Chapters 5 and 6, a number of control strategies may be adopted and a system designer has the prerogative to select the best candidate based on application requirements. Nevertheless, it is important to consider the complexity of the selected control algorithm because the time taken for a compute cycle of this algorithm may be excessively long and thus affect the effectiveness of the control system.

The same control algorithm may require a different computing time because of the software environment on which it runs. For example, the controller operated in the Simulink/MATLAB environment is somewhat slower than a controller converted to an executable binary targeted to run in a real-time environment.

7.3 APPLICATIONS OF RENDERING CONTROL SYSTEMS

The benefits of applying control engineering in real-time computer graphics rendering were mentioned earlier. In this section, a list of application domains will allow readers to understand and appreciate the spectrum of usage with this technology. While this list embodies the broad categories of real-time rendering applications, the technology is certainly not limited by the list.

Computer-aided design and manufacturing (CAD and CAM)—The 3D data sets used in this application domain represent a huge market. The introduction of a control system for such applications will allow users to view data sets even on mobile devices that require little computing power. This application can increase productivity and improve communication when data are moved around.

Computer games and virtual communication—The 3D virtual communication market is growing in the education and corporate services segments. As social networking continues to grow, 3D interactive applications such as games and virtual worlds remain key proponents to online communications. We see the integration of control techniques in real-time rendering as a technology that improves the quality of service of such network communications.

Virtual reality (VR)—These applications aim to create realistic virtual environment that resembles the real world. These applications include product and medical visualisation, scientific uses, military simulation, technical training and support, and 3D sales and marketing software. With increasing demand for higher returns on investments for real-time rendering applications, the technology described in this book can address this requirement by delivering consistent performance in various application settings.

Mixed reality (MR) and rich media—The Internet has evolved into a rich-media communication channel in recent years. Mixed reality (MR) applications blend both virtual and real objects together to create believable and informative worlds. The confluence of these subjects has generated interesting applications and products that require some form of interactive 3D rendering.

A key advantage offered by the control-assisted rendering technology described in this book offers is the deployment of these types of applications over a wide range of computer platforms and human–computer interfaces.

7.3.1 EXTENSION OF CONTROL SYSTEM FRAMEWORK

Control engineering principles have been widely adopted around the world. Control techniques have been adopted across a spectrum of applications including flight dynamics, temperature control, and mechanical systems. We believe the modelling and control system framework described in this book can be extended beyond real-time polygonal rendering (surface shading) to other forms of rendering such as:

Volume rendering [10]
Image-based rendering [61]
Real-time transcoding and compression of video streams [62,63]

7.4 CONVERGENCE WITH FUTURE TECHNOLOGY

Key technology innovation in recent years created several interesting and promising opportunities for control engineering with real-time computer graphics rendering. We provide a summary of the advances and their future prospects below.

7.4.1 GREATER COMPUTING PARALLELISM

The advent of the graphical processing unit (GPU) impelled the quality of real-time computer graphics to progress by leaps and bounds. Today it is typical for a computer to have both a CPU and a GPU dedicated to graphics related task processing. From a control system architectural perspective, this provides a straightforward path to mapping of a controller to the CPU and a plant to the GPU. The benefit of the arrangement is more robust parallel processing and stability in the control system.

7.4.2 INCREASED USE OF MOBILE DEVICES

Decreasing manufacturing costs and sleeker hardware designs flooded the global consumer market with portable and powerful mobile devices. The average time a consumer spends using mobile devices has risen significantly as a result of technology innovations and costs of ownership.

Nevertheless, these devices are still constrained by limited local storage space and less powerful processors than those used in desktop systems. The widespread use of mobile devices provides a tremendous opportunity for the installation of certain adaptive control mechanisms to improve the quality of service of applications. This is an active field of research.

7.4.3 VAST IMPROVEMENTS IN INTERNET INFRASTRUCTURE

As the availability of high-speed Internet connection increases globally, the communication overheads of computers and devices reduce correspondingly. In control system design, this implies significant reductions in latency for closed-loop feedback communications. The improvements represented by Configuration C discussed in Section 7.1 may lead to scalable control system architectures to be implemented across networks and physical locations.

7.5 ECONOMIC AND PRODUCTIVITY IMPACTS

While the technicalities of integrating control engineering with real-time computing have been presented extensively in this book, we feel that the economic and productivity impacts should be emphasised as well. In broad terms and drawing from experiences in industrial fields where control engineering played a significant part, we provide a brief summary below.

7.5.1 ENHANCED PRODUCT LIFESPAN

As data requirements grow rapidly, hardware processing power may not be able to keep pace in many situations. To illustrate, CAD and CAM applications utilise 3D object data extensively. However, an investment in a computer system may yield decreasing productivity as data size scales.

By using adaptive control techniques with level-of-detail management, the work scope of a computer can be expanded significantly, thus prolonging its life as a productive tool. Furthermore, a lengthened product life allows a system to remain useful over a longer period, thus allowing better cost amortisation and lowering the total cost of ownership of computer graphics systems.

7.5.2 INCREASED PRODUCTIVITY

The increasing complexity of designs of many products requires exchanges of design information among various stakeholders in the production pipeline. The advantage of virtual prototyping is that early analysis and insights derived from such activities can help engineers understand the potential pitfalls and test various ideas without incurring the high costs of producing physical prototypes.

In addition to enhanced product lifespan, another benefit is the increased productivity resulting from better use of controlled systems in general. For example, without adaptive real-time rendering, a user would waste precious time generating images. This problem becomes especially acute when real-time visualisation is an integral part of a design and/or manufacturing process. The introduction of a control mechanism can alleviate the display frame rate latency issue and help users become more productive. A direct benefit is a shorter time to market for a product that helps businesses better respond to changing market conditions.

7.5.3 NEW PRODUCTS AND MARKETS

Recent market research and trends indicate that the digital media industry is growing at a phenomenal rate.[*] The forecast remains very positive, driven largely by stronger economies and great demand for digital content around the world. Interactive real-time rendering applications that form a substantial part of digital continue will continue remain relevant for many years.

We believe the technology proposed in this book can lead to many new products that address the needs of various segments of the digital media market. From a socioeconomical perspective, the technology may generate employment and service businesses. It is our hope that adaptive control will serve as a critical component of real-time rendering in the near future.

---

[*] 3D CAD Software Market in the APAC Region 2011–2015. http://www.technavio.com/content/3d-cad-software-market-apac-region-2011-2015. Gartner Forecast: Enterprise Software Markets Worldwide 2008–2013, 1Q09 http://www.gartner.com/DisplayDocument?ref=g_search&id=913424&subref =simplesearch

8 Conclusion

8.1 PERFORMANCE ANALYSIS

In this section, we provide a qualitative and quantitative analysis on the experiment results of previous research in comparison to that from our proposed framework. The analysis is primarily based on three characteristics of the rendering performance—the frame rate stability, transient response and adaptive tracking capability.

Prior to discussing the analysis, it is known generally that the performance of different techniques is best compared by applying them in the same test data-set or environment. However, this cannot be easily accomplished in this research because we are not simply comparing an improvement to an existing technique or algorithm but introducing, establishing and validating a novel rendering architecture. First, the subject matter deals with a 3D rendering approach (polygon-based) which is vastly different compared with other techniques such as image-based and volumetric rendering. This means that the rendering setup and data format cannot be shared or used across the platforms. Second, apart from software configuration certain research spanning interactive 3D rendering techniques surveyed in this book relies on specialized hardware [69,82,85,91] or they work on distributed environments [61,62,67,72] which contrasts greatly with our rendering framework's setup. Therefore we deem the comparison to be adequate by referencing the qualitative and quantitative differences (frame rate stability, transient response and adaptive tracking capability) between the experiment results from previous research and our work.

8.1.1 FRAME RATE STABILITY

One of the key qualitative metric considered in this research which is important in real-time 3D rendering is frame rate stability. A stable frame rate does not only bring about steady visual display that allows positive user experience, it also carries the benefit of optimised resource usage. This can lead to more effective utilisation of the computer's processor cycles compared to a "best-effort" technique that does not guarantee a stable frame rate.

From Figure 8.1, it is evident that Pouderoux and Marvie's technique of streaming 3D terrain data using strip masks [78] did not generate persistently stable frame rates. Gobbetti and Bouvier's multi-resolution technique [88] to control frame rate produces very coarse results as shown in Figure 8.2. The lack of strong adherence to target frame rates is probably most apparent in Jeschke et al.'s [77] research on using imposters as a means to improve frame rates. It is evident from the experiment data as shown in Figure 8.3 that this type of approach is not adaptive in nature and it

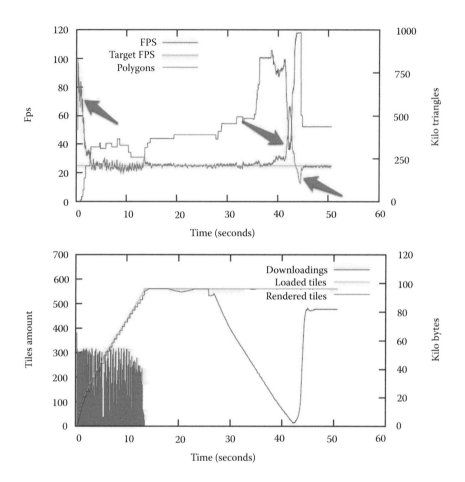

FIGURE 8.1 Experiment results from Pouderoux and Marvie's research [78].

ascribes to the "best-effort" design. Paravati et al.'s [66] adaptive control system did not deliver stable frame rates as well as shown in Figure 8.4 rather the depicted frame rates bear an oscillatory behavior after some steady-state equilibrium.

From a quantitative perspective, all the aforementioned research produced errors in frame rates of more 100% from the target value.

For the purpose of comparison and clarity, we reproduce Figure 5.12 and 5.13 below as Figures 8.5 and 8.6. It can be seen that our proposed modelling and control framework creates absolutely stable frame rates with less than 3% error.

8.1.2 Transient Response

The transient response of a 3D rendering application refers to the quality of its transition as the frame rate changes from one steady-state level to another typically due to changing performance objective. This quality is particularly important at low frame

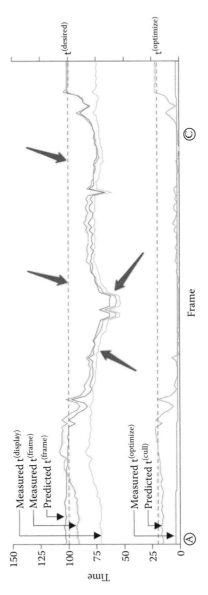

FIGURE 8.2 Experiment results from Gobbetti and Bouvier's multi-resolution technique [88].

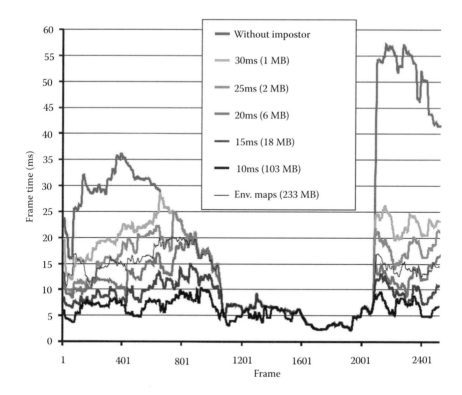

FIGURE 8.3   Experiment results from Jeschke et al.'s approach [77] with usage of imposters.

rate transitions because it can severely affect the user experience due to the "display stuttering" that occurs. With reference to the diagrams in Figure 8.7, Zheng et al. [76] showed in their experiment results that their algorithm in handling distributed rendering produces frame rate accuracies close to the targets. This is however done with bumpy transitions and as depicted in the diagram on the left, there are even oscillations after an initial steady-state.

In contrast to the results shown in Figure 8.7, our fuzzy controller system produces tracking with improved transitions as shown in Figures 5.12, 5.13 (now Figures 8.5 and 8.6) and 6.11 in Chapter 6. There are no sporadic oscillations with large amplitude after the output attains a steady-state level.

8.1.3   ADAPTIVE TRACKING CAPABILITY

While research in interactive 3D rendering purports accurate tracking to a performance objective, what is often not presented is the ability of the technique to adapt to changing performance objectives. We illustrate a practical example where an application that may draw considerable computer hardware processing power can benefit from a longer run-time if the display frame rate can be adaptively changed according

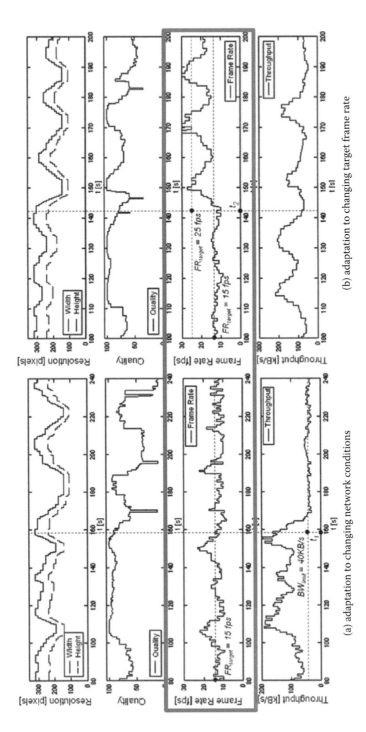

(a) adaptation to changing network conditions

(b) adaptation to changing target frame rate

FIGURE 8.4 Experiment results from Paravati et al's adaptive control technique [66].

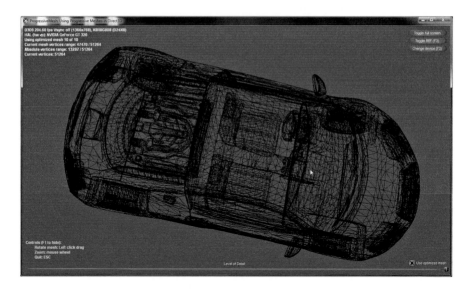

FIGURE 8.5   Screenshot of application in our experiment.

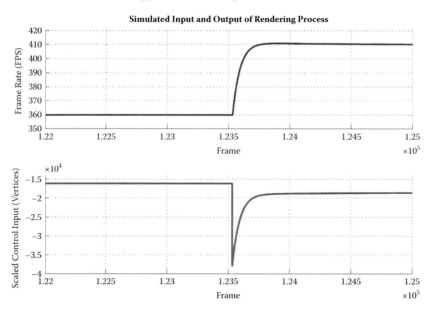

FIGURE 8.6   Reference tracking using PID controller (low to high).

to power levels. For instance, it can start with a default 60 FPS when power is full and change progressively until it reaches 20 FPS at very low power levels. A mechanism like this enhances the usability of the application across a wide operating range but calls for a technique that is robust and flexible enough to support it.

Figure 8.8 shows Li and Shen's research output in time-critical multi-resolution volume rendering [80]. Their algorithm improves the quality of the output which

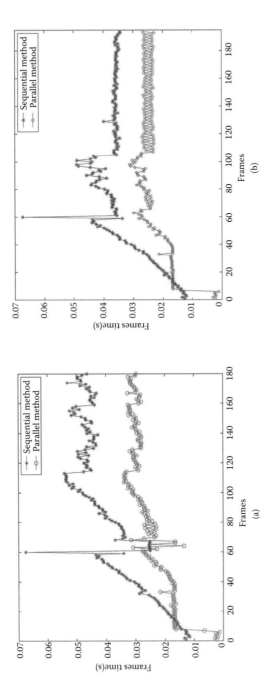

FIGURE 8.7 Experiment results from Zheng et al's work on rendering large 3D models online.

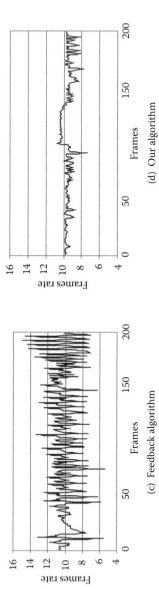

FIGURE 8.8 Experiment results from Li and Shen's research [10] on time-critical multi-resolution volume rendering using 3D texture mapping hardware.

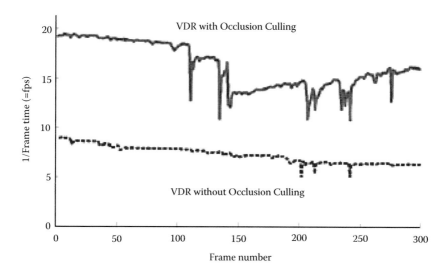

FIGURE 8.9 Quick-VDR: Interactive View-Dependent Rendering of Massive Models, Yoon et al. [81].

uses the feedback algorithm. However, it only presents to the reader a single frame rate objective.

Similarly, the experiment results from Yoon et al's work [81] on interactive view-dependent rendering of massive models show improved frame rates but only with oscillations and unstable frame rates. There is no depiction in Figure 8.9 of any adaptive capability in handling discrete frame rate level changes.

Another example that shows frame rate improvement is found in Figure 8.10 from Scherzer, Yang, and Mattausch's research [69] on exploiting temporal coherence in real-time rendering. The output from their technique is compared against other approaches but there is no information on the technique's handling of frame rate level transitions and the corresponding transient response.

Our proposed control system for 3D rendering however produces fast and direct transitions from one steady-state transition to another as shown in Figure 6.12 in Chapter 6. The small amount of delay in the tracking is due to the implementation of the controller and plant which involves network communication.

8.2   SUMMARY

In this book, we described an intelligent real-time rendering system based on our research in the fields of control engineering, system identification, and real-time computer graphics. We introduced a novel control system framework using a closed-loop feedback design with the rendering process—the plant to be controlled. The salient areas of this research were the detailed process steps for deriving system models for the real-time rendering process. The techniques were not discussed in previous research. We devised models that can capture both linear and non-linear characteristics of the rendering process.

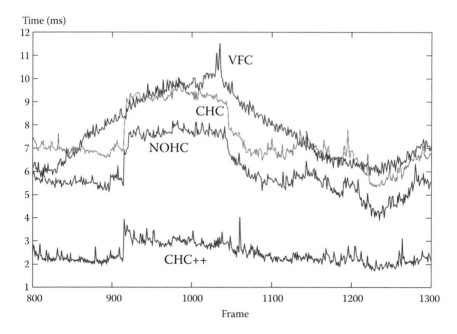

FIGURE 8.10 Experiment results from Scherzer, Yang, and Mattausch's [69] research on exploiting temporal coherence in real-time rendering.

Furthermore, we investigated and developed control frameworks using model-based and model-less approaches for real-time rendering. The frameworks discussed apply to conventional PID controls for linear processes, piecewise linear controls, and the use of soft computing techniques such as neuro-fuzzy control for setting up systems without formal model definitions.

Our experiments show that it is possible to model real-time rendering accurately. We performed further experiments that validated the performance of our control system utilising both PID and fuzzy controllers in different arrangements.

8.3 FUTURE WORK

We hope our research will inspire appreciation for and wider adoption of theoretic controls for computer graphics applications. Based on the initial objectives we set and met in our research, we hope that future work that may generate interest in some related topics.

First, a global geometry manager for rendering software would be desirable for handling 3D scenes involving many different objects. More specifically, a managing device could resemble the common hierarchical scene graph by which objects and sub-objects are organised in a contextually meaningful manner. The function of this geometry manager is to provide better resolution control for the geometry load of a 3D scene by determining which objects must scaled appropriately based on their geometric constructs.

Certain visual-based criteria may serve as guidelines determining the extent to which individual 3D objects should be scaled. In other words, the manager will act as a "middle man" between the control action and the rendering process, distributing the effects of the control action in an elegant way to change individual objects within the 3D scene. This would extend the utility of the proposed control system framework to a larger pool of applications.

The next possible extension to this research is investigating other input variables to the plant that may be used in the control system framework. Although it may be difficult to find many variables that are easily accessible from the rendering pipeline and able to be changed at reasonable resolution rates, it may be possible to introduce certain user-defined input parameters specific to certain applications. For example instead of computer graphics rendering pipeline inputs, the control system framework may include hardware-related resources such as memory and CPU utilisation that can impact rendering process performance. Also, recent advancements in computer graphics hardware and techniques may expose new input variables that may be considered in future control system frameworks.

The experiments we conducted focused largely on verification of the key concepts of introducing control principles in real-time computer graphics rendering. As a result, the simulation environment construct was performed using available test software. For practical applications, generic libraries should be developed so that integration into different real-time rendering software can be achieved easily. However, this proposal can involve significant time for code verification and optimisation of runtime efficiency.

Annex A: Sample Applications

A.1 OVERVIEW

The functionalities of the two sample applications used in the experiments described in this book are detailed below. They were used as the main rendering processes that were modified for implementation of a closed-loop feedback control system.

A.2 PROGRESSIVEMESH SAMPLE[*]

This ProgressiveMesh sample shows how an application can use the D3DX progressive mesh functionality to simplify meshes for faster rendering. It is a specialised mesh object that can increase or decrease its geometry complexity, thereby providing flexibility when drawing a mesh so that performance can be maintained at a steady level. This feature is useful when providing LoD support. (Note: The simplification algorithm used to generate progressive meshes is based on Hugues Hoppe's Siggraph papers.)

A.3 HOW SAMPLE WORKS

The functionalities of progressive meshes are provided by ID3DXPMesh. The mesh interface is similar to ID3DXMesh with additional methods for managing complexity. To generate a progressive mesh, call D3DXGeneratePMesh. The progressive mesh can be used just like a regular mesh. To render it, a sample loops through its materials and calls *ID3DXBaseMesh::DrawSubset* to send the geometry subset to the device. To adjust the level of detail (LoD) of the progressive mesh, the sample calls *ID3DXPMesh::SetNumVertices* and passes it the desired number of vertices. A progressive mesh will simplify or enhance its geometry to match the number of vertices as closely as possible.

The sample also shows an optimisation technique for progressive meshes by trimming multiple meshes. Trimming limits the maximum and minimum number of vertices or faces a progressive mesh can have. The sample divides the range (maximum to minimum) of the progressive mesh vertices into ten sub-ranges. After the sub-ranges are computed, the sample creates ten progressive meshes by calling *ID3DXPMesh::ClonePMeshFVF* on the original mesh. Then the sample *calls ID3DXPMesh::TrimByVertices* on each progressive mesh using a different sub-range.

After setting the range of vertices, the sample calls *ID3DXPMesh::Optimize-BaseLOD* to optimise the mesh vertex and index buffers. When a user changes the vertex count, the new vertex count is checked against the range of the optimised progressive mesh set, and the mesh whose range contains the desired vertex count is selected by calling *ID3DXPMesh::SetNumVertices*.

---

[*] Documentation reproduced from Microsoft DirectX SDK 2009.

The advantage of multiple meshes over a single mesh is that adjusting LoD is more efficient. The performance load of changing LoD is directly proportional to the difference in complexity (represented by vertex count in this sample). Simplifying a mesh by reducing the number of vertices by 10 takes less time than reducing the vertex count by 100. That is why this sample achieves better performance by trimming several progressive meshes, each of which covers a smaller LoD range.

A.4   TESSELLATION SAMPLE*

This sample demonstrates mesh tessellation in Microsoft Direct3D. Mesh tessellation subdivides mesh triangles to yield a mesh with finer geometry details that produces better results even with per-vertex lighting. Mesh tessellation is often used to implement LoD where meshes closer to the viewer are rendered with more details and more distant meshes are rendered with less detail.

A.5   HOW SAMPLE WORKS

The sample can run in one of two modes: hardware or software tessellation. The user can set the tessellation level to different values and see how the mesh changes in reaction to the level adjustment.

When running in hardware tessellation mode, the sample tessellates the mesh by setting the *Device.NPatchMode* property that sets the number of tessellation segments into which the device will tessellate each mesh segment. For instance, specifying 3.0 will cause each original segment in the input mesh to be tessellated into three segments. Tessellation happens in real-time, after the mesh draw calls in the render loop.

When running in software tessellation mode, the sample does not rely on the hardware to handle on-the-fly tessellation. The sample must process the mesh and obtain the desired detail level before rendering it. The code achieves this by calling *Mesh.TessellateNPatches* to take an input mesh and a segment count and then output another mesh that represents the tessellated version of the input mesh. The sample can then render this tessellated mesh using any standard mechanism.

A.6   SAMPLES

```
//————————————————————————————————————————————————
//Program modified by Gabriyel Wong from EnhancedMesh example from
//Microsoft DirectX 9 SDK
//Additional components added: network communication, tessellation and
//rendering quality controls.
//Author: Gabriyel Wong
//Original code copyrights Microsoft.
//————————————————————————————————————————————————
```

---

* Documentation reproduced from Microsoft DirectX SDK 2009.

```
#include "DXUT.h"
#include "DXUTcamera.h"
#include "DXUTsettingsdlg.h"
#include "SDKmisc.h"
#include "resource.h"
#include "skybox.h"
#include "XNet.h"
#include <iostream>
#include <fstream>
#include <time.h>


//#define DEBUG_VS //Uncomment this line to debug vertex shaders
//#define DEBUG_PS //Uncomment this line to debug pixel shaders

#define NUM_TONEMAP_TEXTURES    5       //Number of stages in the 3x3
                                        //down-scaling of average luminance
                                        //textures
#define NUM_BLOOM_TEXTURES      2
#define RGB16_MAX               100

enum ENCODING_MODE
{
   FP16,
   FP32,
   RGB16,
   RGBE8,
   NUM_ENCODING_MODES
};

enum RENDER_MODE
{
   DECODED,
   RGB_ENCODED,
   ALPHA_ENCODED,
   NUM_RENDER_MODES
};

struct TECH_HANDLES
{
   D3DXHANDLE XRay;
   D3DXHANDLE SimpleLighting;
   D3DXHANDLE SpecularLighting;
   D3DXHANDLE ToonEffect;
   D3DXHANDLE Reflect;
   D3DXHANDLE ReflectSpecular;
};

struct SCREEN_VERTEX
{
   D3DXVECTOR4 pos;
   D3DXVECTOR2 tex;
   static const DWORD FVF;
};

const DWORD SCREEN_VERTEX::FVF = D3DFVF_XYZRHW | D3DFVF_TEX1;
```

```
//————————————————————————————————————————————————————
//Global variables
//————————————————————————————————————————————————————
IDirect3DDevice9*      g_pd3dDevice;       //Direct3D device
LPCWSTR                g_Models[6] = { L"Models\\dragon.x",
                                        L"Models\\engine.x",
                                        L"Models\\audi.x",
                                        L"Models\\skull.x"};
DWORD      g_ModelCount = 6;
DWORD      g_DwShaderFlags = 0;
DWORD      g_CurrentModel = 0;
CSkybox    g_Skybox;
ID3DXFont*             g_pFont = NULL;        //Font for drawing text
ID3DXSprite*           g_pTextSprite = NULL; //Sprite for batching draw
                                             text calls
ID3DXEffect*           g_pEffect = NULL;     //D3DX effect interface
CModelViewerCamera     g_Camera;             //A model viewing camera
IDirect3DTexture9*     g_pDefaultTex = NULL; //Default texture for
                                             texture-less material
bool                   g_bShowHelp = true;   //If true, it renders the UI
                                             control text
CDXUTDialogResourceManager g_DialogResourceManager;//manager for
                                             shared resources of dialogs
CD3DSettingsDlg        g_SettingsDlg;        //Device settings dialog
CDXUTDialog            g_HUD;                //dialog for standard
                                             controls
CDXUTDialog            g_SampleUI;           //dialog for sample specific
                                             controls
ID3DXMesh*             g_pMeshSysMem = NULL; //system memory version of
                                             mesh, lives through
                                             resise's
ID3DXMesh*             g_pMeshEnhanced = NULL; //vid mem version of mesh
                                             that is enhanced
UINT                   g_dwNumSegs = 1;      //number of segments per edge
                                             (tesselation level)
D3DXMATERIAL*          g_pMaterials = NULL;  //pointer to material info in
                                             m_pbufMaterials
LPDIRECT3DTEXTURE9*    g_ppTextures = NULL;  //array of textures, entries
                                             are NULL if no texture
                                             specified
DWORD                  g_dwNumMaterials = NULL; //number of materials
D3DXVECTOR3            g_vObjectCenter;      //Center of bounding sphere
                                             of object
FLOAT                  g_fObjectRadius;      //Radius of bounding sphere
                                             of object
D3DXMATRIXA16          g_mCenterWorld;       //World matrix to center the
                                             mesh
ID3DXBuffer*           g_pbufMaterials = NULL; //contains both the
                                             materials data and the
                                             filename strings
ID3DXBuffer*           g_pbufAdjacency = NULL; //Contains the adjacency
                                             info loaded with the mesh
bool                   g_bUseHWNPatches = true;
bool                   g_bWireframe = false;
PDIRECT3DSURFACE9      g_pMSRT = NULL;       //Multi-Sample float render
                                             target
```

```
PDIRECT3DSURFACE9   g_pMSDS = NULL;        //Depth Stencil surface for
                                             the float RT
LPDIRECT3DTEXTURE9  g_pTexRender;          //Render target texture
LPDIRECT3DTEXTURE9  g_pTexBrightPass;      //Bright pass filter
LPD3DXMESH          g_pMesh;
LPDIRECT3DTEXTURE9  g_apTexToneMap[NUM_TONEMAP_TEXTURES]; //Tone
                                             mapping calculation
                                             textures
LPDIRECT3DTEXTURE9  g_apTexBloom[NUM_BLOOM_TEXTURES]; //Blooming
                                             effect intermediate texture
bool                g_bBloom;              //Bloom effect on/off
ENCODING_MODE       g_eEncodingMode;
RENDER_MODE         g_eRenderMode;
TECH_HANDLES        g_aTechHandles;
TECH_HANDLES*       g_pCurTechnique;
bool                g_bShowText;
double              g_aPowsOfTwo[257];     //Lookup table for log
                                             calculations
bool                g_bSupportsR16F = false;
bool                g_bSupportsR32F = false;
bool                g_bSupportsD16 = false;
bool                g_bSupportsD32 = false;
bool                g_bSupportsD24X8 = false;
bool                g_bUseMultiSample = false; //True when using
                                             multisampling on a
                                             supported back buffer
D3DMULTISAMPLE_TYPE g_MaxMultiSampleType = D3DMULTISAMPLE_NONE;
                                         //Non-Zero when g_
                                           bUseMultiSample is true
DWORD               g_dwMultiSampleQuality = 0;
                                         //Used when we have
                                           multisampling on a
                                           backbuffer
IDirect3DCubeTexture9* g_pCubeTexture = NULL;
int       g_CurrentCubeTexture = 1;
LPCWSTR   g_CubeTextures[16] = { L"Light Probes\\street.dds", L"Street",
                                 L"Light Probes\\castle.dds", L"Castle",
                                 L"Light Probes\\park.dds", L"Park",
                                 L"Light Probes\\night.dds", L"Night",
                                 //L"Light Probes\\ParkLow.dds", L"Park
                                 Low",
                                 //L"Light Probes\\Park.dds", L"Park
                                 High",
                                 //L"Light Probes\\CreekLow.dds",
                                 L"Creek Low",
                                 //L"Light Probes\\VasaLow.dds", L"Vasa
                                 Low",
                                 //L"Light Probes\\Vasa.dds", L"Vasa
                                 High"
};

float     g_fModelReflectivity = 0.75f;
CHAR*     g_Techniques[] = {"XRay", "SimpleLighting",
"SpecularLighting", "ToonEffect", "Reflect", "ReflectSpecular",};
LPCWSTR   g_TechniqueNames[] = {L"X-Ray", L"Diffuse Lighting",
L"Specular Lighting", L"Toon Effect",
L"Reflection + Diffuse", L"Reflection + Specular"};
```

```
int  g_CurrentTechnique = 0;
int  g_OriginalNumFaces = 0;

//————————————————————————————————————————————————————————————
//UI control IDs
//————————————————————————————————————————————————————————————
#define IDC_TOGGLEFULLSCREEN          1
#define IDC_TOGGLEREF                 3
#define IDC_CHANGEDEVICE              4
#define IDC_FILLMODE                  5
#define IDC_SEGMENTLABEL              6
#define IDC_SEGMENT                   7
#define IDC_HWNPATCHES                8
#define IDC_CUBETEXTURELABEL          9
#define IDC_CUBETEXTURE               10
#define IDC_MODELREFLECTIVITYLABEL    11
#define IDC_MODELREFLECTIVITY         12
#define IDC_ACTIVEEFFECTLABEL         13
#define IDC_ACTIVEEFFECT              14


//————————————————————————————————————————————————————————————
//Forward declarations
//————————————————————————————————————————————————————————————
bool CALLBACK IsDeviceAcceptable(D3DCAPS9* pCaps, D3DFORMAT
AdapterFormat, D3DFORMAT BackBufferFormat, bool bWindowed,
void* pUserContext);
bool CALLBACK ModifyDeviceSettings(DXUTDeviceSettings*
pDeviceSettings, void* pUserContext);
HRESULT CALLBACK OnCreateDevice(IDirect3DDevice9* pd3dDevice, const
D3DSURFACE_DESC* pBackBufferSurfaceDesc,
void* pUserContext);
HRESULT CALLBACK OnResetDevice(IDirect3DDevice9* pd3dDevice, const
D3DSURFACE_DESC* pBackBufferSurfaceDesc,
void* pUserContext);
void CALLBACK OnFrameMove(double fTime, float fElapsedTime, void*
pUserContext);
void CALLBACK OnFrameRender(IDirect3DDevice9* pd3dDevice, double
fTime, float fElapsedTime, void* pUserContext);
LRESULT CALLBACK MsgProc(HWND hWnd, UINT uMsg, WPARAM wParam, LPARAM
lParam, bool* pbNoFurtherProcessing, void* pUserContext);
void CALLBACK KeyboardProc(UINT nChar, bool bKeyDown, bool bAltDown,
void* pUserContext);
void CALLBACK OnGUIEvent(UINT nEvent, int nControlID, CDXUTControl*
pControl, void* pUserContext);
void CALLBACK OnLostDevice(void* pUserContext);
void CALLBACK OnDestroyDevice(void* pUserContext);

void InitApp();
HRESULT LoadMesh(IDirect3DDevice9* pd3dDevice, WCHAR* strFileName,
ID3DXMesh** ppMesh);
void RenderText();
HRESULT GenerateEnhancedMesh(IDirect3DDevice9* pd3dDevice, UINT
cNewNumSegs);

//Use compile symbol NETWORK_CONTROL to set network configuration
#ifdef NETWORK_CONTROL
```

```
//Threading info
#define MAX_THREADS 2
#define BUF_SIZE 255


DWORD WINAPI SendDataThreadFunction(LPVOID lpParam);
DWORD WINAPI ReceiveDataThreadFunction(LPVOID lpParam);

void ErrorHandler(LPTSTR lpszFunction);

//Sample custom data structure for threads to use.
//This is passed by void pointer so it can be any data type
//that can be passed using a single void pointer (LPVOID).
typedef struct MyData {
    int val1;
    int val2;
} MYDATA, *PMYDATA;

MyDataStruct ds, dr;

DWORD WINAPI SendDataThreadFunction(LPVOID lpParam)
{
     XNet* xnet = new XNet();
     xnet->init(CLIENT, 64000, UDP, "localhost");
     while(1)  //Keep the thread alive
     {
        xnet->sendData(ds);
        printf("SendDataThreadFunction:%f\n", ds.data[0]);
     }
   return 0;
}


DWORD WINAPI ReceiveDataThreadFunction(LPVOID lpParam)
{
     XNet* xnet = new XNet();
     xnet->init(SERVER, 64001, UDP);
     while(1)  //Keep the thread alive
     {
        dr.data[0] = xnet->receiveData().data[0];
        dr.data[1] = xnet->receiveData().data[1];
     }
   return 0;
}


PMYDATA   pDataArray[MAX_THREADS];
DWORD     dwThreadIdArray[MAX_THREADS];
HANDLE    hThreadArray[MAX_THREADS];
#endif

std::ofstream logfile;
//————————————————————————————————————————————
//Entry point to the program. Initialises everything and goes into a
//message processing loop. Idle time is used to render the scene.
//————————————————————————————————————————————
int main(void)
{
   logfile.open ("data.log");
   long startTime = time(NULL);
```

```
#ifdef NETWORK_CONTROL
    //Since rendering application is C++, and the system is SISO/MISO,
    // there is only one data channel to send back to controller, i.e. y.
    //Note data type and container size.
    ds.data[0] = 999;    //Initialization value
    ds.data[1] = 999;

    //Clear receive buffer
    dr.data[0] = 999;
    dr.data[1] = 999;

    for(int i = 0; i<MAX_THREADS; i++)
    {
        //Allocate memory for thread data.
        pDataArray[i] = (PMYDATA) HeapAlloc(GetProcessHeap(),
HEAP_ZERO_MEMORY, sizeof(MYDATA));

        if(pDataArray[i] = = NULL)
        {
            //If the array allocation fails, the system is out of memory
            //so there is no point in trying to print an error message.
            //Just terminate execution.
                ExitProcess(2);
        }

        //0 - Send, 1 - Receive
        if (i = = 0)
        {
            hThreadArray[i] = CreateThread(
                NULL,                      //default security attributes
                0,                         //use default stack size
                SendDataThreadFunction,    //thread function name
                pDataArray[i],             //argument to thread function
                0,                         //use default creation flags
                &dwThreadIdArray[i]);      //returns the thread identifier
        }

        if (i = = 1)
        {
            hThreadArray[1] = CreateThread(
                NULL,                        //default security attributes
                0,                           //use default stack size
                ReceiveDataThreadFunction,   //thread function name
                pDataArray[1],               //argument to thread function
                0,                           //use default creation flags
                &dwThreadIdArray[1]);        //returns the thread identifier
        }

        //Check the return value for success.
        //If CreateThread fails, terminate execution.
        //This will automatically clean up threads and memory.

        if (hThreadArray[i] = = NULL)
        {
            printf("Error creating thread...!\n");
            ExitProcess(3);
        }
```

```
   }//End of main thread creation loop.

   //Wait until all threads have terminated.
   WaitForMultipleObjects(MAX_THREADS, hThreadArray, TRUE, INFINITE);

   //Close all thread handles and free memory allocations.
   for(int i = 0; i<MAX_THREADS; i++)
   {
      CloseHandle(hThreadArray[i]);
      if(pDataArray[i] ! = NULL)
      {
         HeapFree(GetProcessHeap(), 0, pDataArray[i]);
         pDataArray[i] = NULL; //Ensure address is not reused.
      }
   }
#endif

   //Enable run-time memory check for debug builds.
#if defined(DEBUG) | defined(_DEBUG)
   _CrtSetDbgFlag(_CRTDBG_ALLOC_MEM_DF | _CRTDBG_LEAK_CHECK_DF);
#endif

   //Set the callback functions
   DXUTSetCallbackD3D9DeviceAcceptable(IsDeviceAcceptable);
   DXUTSetCallbackD3D9DeviceCreated(OnCreateDevice);
   DXUTSetCallbackD3D9DeviceReset(OnResetDevice);
   DXUTSetCallbackD3D9FrameRender(OnFrameRender);
   DXUTSetCallbackD3D9DeviceLost(OnLostDevice);
   DXUTSetCallbackD3D9DeviceDestroyed(OnDestroyDevice);
   DXUTSetCallbackMsgProc(MsgProc);
   DXUTSetCallbackKeyboard(KeyboardProc);
   DXUTSetCallbackFrameMove(OnFrameMove);
   DXUTSetCallbackDeviceChanging(ModifyDeviceSettings);

   //Initialize DXUT and create the desired Win32 window and Direct3D
   //device for the application
   DXUTSetCursorSettings(true, true);//Show the cursor and clip it when
in full screen
   InitApp();
   DXUTInit(true, true);//Parse the command line and show msgboxes
   DXUTSetHotkeyHandling(true, true, true); //handle the default
hotkeys
   DXUTCreateWindow(L"Enhanced Mesh - N-Patches");
   DXUTCreateDevice(true, 1366, 768);
   DXUTMainLoop();

   //Perform any application-level cleanup here. Direct3D device
   //resources are released within the appropriate callback functions
   //and therefore don't require any cleanup code here.

      logfile << "Duration(sec): " << time(NULL) - startTime << std::endl;
      logfile.close();
   return DXUTGetExitCode();
}
```

```
//————————————————————————————————————————————————
//Initialize the app
//————————————————————————————————————————————————
void InitApp()
{
   g_pFont = NULL;
   g_pEffect = NULL;
   g_bShowHelp = true;
   g_bShowText = true;

   g_pMesh = NULL;
   g_pTexRender = NULL;

   g_bBloom = TRUE;
   g_eEncodingMode = RGBE8;
   g_eRenderMode = DECODED;

   g_pCurTechnique = &g_aTechHandles;

   for(int i = 0; i < = 256; i++)
   {
      g_aPowsOfTwo[i] = powf(2.0f, (float)(i - 128));
   }

   ZeroMemory(g_apTexToneMap, sizeof(g_apTexToneMap));
   ZeroMemory(g_apTexBloom, sizeof(g_apTexBloom));
   //ZeroMemory(g_aTechHandles, sizeof(g_aTechHandles));

   //Initialize dialogs
   g_SettingsDlg.Init(&g_DialogResourceManager);
   g_HUD.Init(&g_DialogResourceManager);
   g_SampleUI.Init(&g_DialogResourceManager);

   g_HUD.SetCallback(OnGUIEvent); int iY = 10;
   g_HUD.AddButton(IDC_TOGGLEFULLSCREEN, L"Toggle full screen", 35, iY,
125, 22);
   g_HUD.AddButton(IDC_TOGGLEREF, L"Toggle REF (F3)", 35, iY + = 24,
125, 22);
   g_HUD.AddButton(IDC_CHANGEDEVICE, L"Change device (F2)", 35, iY + =
24, 125, 22, VK_F2);

   g_SampleUI.SetCallback(OnGUIEvent); iY = 10;
   g_SampleUI.AddComboBox(IDC_FILLMODE, 10, iY, 150, 24, L'F');
   g_SampleUI.GetComboBox(IDC_FILLMODE)->AddItem(L"(F)illmode: Solid",
(void*)0);
   g_SampleUI.GetComboBox(IDC_FILLMODE)->AddItem(L"(F)illmode:
Wireframe", (void*)1);
   g_SampleUI.AddStatic(IDC_SEGMENTLABEL, L"Number of segments: 1", 10,
iY + = 30, 150, 16);
   g_SampleUI.AddSlider(IDC_SEGMENT, 10, iY + = 14, 150, 24, 1, 10, 1);
   g_SampleUI.AddCheckBox(IDC_HWNPATCHES, L"Use hardware N-patches",
10, iY + = 26, 150, 20, true, L'H');

      g_SampleUI.AddStatic(IDC_CUBETEXTURELABEL, L"Skymap Texture:",
10, iY + = 26, 150, 16);
      g_SampleUI.AddComboBox(IDC_CUBETEXTURE, 10, iY + = 14, 150, 24);
```

```
    for(int i = 1; i < 16; i + = 2)
    {
        g_SampleUI.GetComboBox(IDC_CUBETEXTURE)->
AddItem(g_CubeTextures[i], (void*)i);
    }

    g_SampleUI.AddStatic(IDC_MODELREFLECTIVITYLABEL, L"Model
Reflectivity : 30", 10, iY + = 35, 150, 16);
    g_SampleUI.AddSlider(IDC_MODELREFLECTIVITY, 10, iY + = 14, 150,
24, 0, 100, 30);
    g_SampleUI.AddStatic(IDC_ACTIVEEFFECTLABEL, L"Active Shader", 10,
iY + = 26, 150, 16);
    g_SampleUI.AddComboBox(IDC_ACTIVEEFFECT, -10, iY + = 14, 170, 24);
    for(int i = 0; i < 6; ++i)
    {
        g_SampleUI.GetComboBox(IDC_ACTIVEEFFECT)->
AddItem(g_TechniqueNames[i], (void*)i);
    }
}

//————————————————————————————————————————————————
//Rejects any D3D9 devices that aren't acceptable to the app by
//returning false
//————————————————————————————————————————————————
bool CALLBACK IsDeviceAcceptable(D3DCAPS9* pCaps, D3DFORMAT
AdapterFormat, D3DFORMAT BackBufferFormat, bool bWindowed, void*
pUserContext)
{
    //Skip backbuffer formats that don't support alpha blending
    IDirect3D9* pD3D = DXUTGetD3D9Object();
    if(FAILED(pD3D->CheckDeviceFormat(pCaps->AdapterOrdinal,
pCaps->DeviceType, AdapterFormat, D3DUSAGE_QUERY_POSTPIXELSHADER_
BLENDING, D3DRTYPE_TEXTURE, BackBufferFormat)))
        return false;

    //Must support pixel shader 2.0
    if(pCaps->PixelShaderVersion < D3DPS_VERSION(2, 0))
        return false;

    return true;
}

//————————————————————————————————————————————————
//Before a device is created, modify the device settings as needed
//————————————————————————————————————————————————
bool CALLBACK ModifyDeviceSettings(DXUTDeviceSettings*
pDeviceSettings, void* pUserContext)
{
    assert(DXUT_D3D9_DEVICE = = pDeviceSettings->ver);

    HRESULT hr;
    IDirect3D9* pD3D = DXUTGetD3D9Object();
    D3DCAPS9 caps;

    V(pD3D->GetDeviceCaps(pDeviceSettings->d3d9.AdapterOrdinal,
pDeviceSettings->d3d9.DeviceType, &caps));
```

```
   //Turn vsync off
   pDeviceSettings->d3d9.pp.PresentationInterval =
D3DPRESENT_INTERVAL_IMMEDIATE;
   g_SettingsDlg.GetDialogControl()->
GetComboBox(DXUTSETTINGSDLG_PRESENT_INTERVAL)->SetEnabled(false);

   //If device doesn't support HW T&L or doesn't support 1.1 vertex
   //shaders in HW then switch to SWVP.
   if((caps.DevCaps & D3DDEVCAPS_HWTRANSFORMANDLIGHT) = = 0 ||
      caps.VertexShaderVersion < D3DVS_VERSION(1, 1))
   {
      pDeviceSettings->d3d9.BehaviourFlags =
D3DCREATE_SOFTWARE_VERTEXPROCESSING;
   }

   //Debugging vertex shaders requires either REF or software vertex
   //processing and debugging pixel shaders requires REF.
#ifdef DEBUG_VS
   if(pDeviceSettings->d3d9.DeviceType ! = D3DDEVTYPE_REF)
   {
      pDeviceSettings->d3d9.BehaviourFlags & =
~D3DCREATE_HARDWARE_VERTEXPROCESSING;
      pDeviceSettings->d3d9.BehaviourFlags & = ~D3DCREATE_PUREDEVICE;
      pDeviceSettings->d3d9.BehaviourFlags | =
D3DCREATE_SOFTWARE_VERTEXPROCESSING;
   }
#endif
#ifdef DEBUG_PS
   pDeviceSettings->d3d9.DeviceType = D3DDEVTYPE_REF;
#endif
   //For the first device created if its a REF device, optionally
   //display a warning dialog box
   static bool s_bFirstTime = true;
   if(s_bFirstTime)
   {
      s_bFirstTime = false;
      if(pDeviceSettings->d3d9.DeviceType = = D3DDEVTYPE_REF)
         DXUTDisplaySwitchingToREFWarning(pDeviceSettings->ver);
   }

   return true;
}

//————————————————————————————————————————————————————————
//Generate a mesh that can be tesselated.
//————————————————————————————————————————————————————————
HRESULT GenerateEnhancedMesh(IDirect3DDevice9* pd3dDevice, UINT
dwNewNumSegs)
{
   LPD3DXMESH pMeshEnhancedSysMem = NULL;
   LPD3DXMESH pMeshTemp;
   HRESULT hr;

   if(g_pMeshSysMem = = NULL)
      return S_OK;
```

```
   //if using hw, just copy the mesh
   if(g_bUseHWNPatches)
   {
      hr = g_pMeshSysMem->CloneMeshFVF(D3DXMESH_WRITEONLY |
D3DXMESH_NPATCHES | (g_pMeshSysMem->GetOptions() & D3DXMESH_32BIT),
g_pMeshSysMem->GetFVF(), pd3dDevice, &pMeshTemp);
      if(FAILED(hr))
         return hr;
   }
   else //Tesselate the mesh in software
   {
      //Create an enhanced version of the mesh, will be in sysmem since
      //source is
      hr = D3DXTessellateNPatches(g_pMeshSysMem, (DWORD*)g_
pbufAdjacency->GetBufferPointer(), (float)dwNewNumSegs, FALSE,
&pMeshEnhancedSysMem, NULL);
      if(FAILED(hr))
      {
         //If the tessellate failed, there might have been more
         //triangles or vertices than can fit into a 16bit mesh,
         //so try cloning to 32bit before tessellation

         hr = g_pMeshSysMem->CloneMeshFVF(D3DXMESH_SYSTEMMEM |
D3DXMESH_32BIT, g_pMeshSysMem->GetFVF(), pd3dDevice, &pMeshTemp);
         if(FAILED(hr))
            return hr;

         hr = D3DXTessellateNPatches(pMeshTemp,
(DWORD*)g_pbufAdjacency->GetBufferPointer(), (float)dwNewNumSegs, FALSE,
&pMeshEnhancedSysMem, NULL);
         if(FAILED(hr))
         {
            pMeshTemp->Release();
            return hr;
         }
         pMeshTemp->Release();
      }

      //Make a video memory version of the mesh
      //Only set WRITEONLY if it doesn't use 32bit indices, because
      //those often need to be emulated, which means that D3DX needs
      //read-access.
      DWORD dwMeshEnhancedFlags = pMeshEnhancedSysMem->GetOptions() &
D3DXMESH_32BIT;
      if((dwMeshEnhancedFlags & D3DXMESH_32BIT) = = 0)
         dwMeshEnhancedFlags | = D3DXMESH_WRITEONLY;
      hr = pMeshEnhancedSysMem->CloneMeshFVF(dwMeshEnhancedFlags,
g_pMeshSysMem->GetFVF(), pd3dDevice, &pMeshTemp);
      if(FAILED(hr))
      {
         SAFE_RELEASE(pMeshEnhancedSysMem);
         return hr;
      }

      //Latch in the enhanced mesh
      SAFE_RELEASE(pMeshEnhancedSysMem);
   }
```

```
   SAFE_RELEASE(g_pMeshEnhanced);
   g_pMeshEnhanced = pMeshTemp;
   g_dwNumSegs = dwNewNumSegs;

   return S_OK;
}

HRESULT LoadEffect(IDirect3DDevice9* pd3dDevice, LPCWSTR effect)
{
     HRESULT hr;

     SAFE_RELEASE(g_pEffect);

     //Read the D3DX effect file
   WCHAR str[MAX_PATH];
   V_RETURN(DXUTFindDXSDKMediaFileCch(str, MAX_PATH, effect));

   //If this fails, there should be debug output as to
   //they the.fx file failed to compile
   V_RETURN(D3DXCreateEffectFromFile(pd3dDevice, str, NULL, NULL,
g_DwShaderFlags, NULL, &g_pEffect, NULL));
     return S_OK;
}

HRESULT LoadMesh(IDirect3DDevice9* pd3dDevice, DWORD meshIndex)
{
   HRESULT hr;
   WCHAR wszMeshDir[MAX_PATH];
   WCHAR wszWorkingDir[MAX_PATH];
   IDirect3DVertexBuffer9* pVB = NULL;

   for(UINT i = 0; i < g_dwNumMaterials; i++)
      SAFE_RELEASE(g_ppTextures[i]);
   SAFE_DELETE_ARRAY(g_ppTextures);
   SAFE_RELEASE(g_pMeshSysMem);
   SAFE_RELEASE(g_pbufMaterials);
   SAFE_RELEASE(g_pbufAdjacency);
      SAFE_RELEASE(g_pMeshEnhanced);

   //Load the mesh
      V_RETURN(DXUTFindDXSDKMediaFileCch(wszMeshDir, MAX_PATH,
g_Models[meshIndex]));
    V_RETURN(D3DXLoadMeshFromX(wszMeshDir, D3DXMESH_SYSTEMMEM,
pd3dDevice, &g_pbufAdjacency, &g_pbufMaterials, NULL,
&g_dwNumMaterials, &g_pMeshSysMem));
      g_OriginalNumFaces = g_pMeshSysMem->GetNumFaces();
   //Initialize the mesh directory string
   WCHAR* pwszLastBSlash = wcsrchr(wszMeshDir, L'\\');
   if(pwszLastBSlash)
      *pwszLastBSlash = L'\0';
   else
      StringCchCopyW(wszMeshDir, MAX_PATH, L".");

   //Lock the vertex buffer, to generate a simple bounding sphere
   hr = g_pMeshSysMem->GetVertexBuffer(&pVB);
   if(FAILED(hr))
      return hr;
```

```
   void* pVertices = NULL;
   hr = pVB->Lock(0, 0, &pVertices, 0);
   if(FAILED(hr))
   {
      SAFE_RELEASE(pVB);
      return hr;
   }

   hr = D3DXComputeBoundingSphere((D3DXVECTOR3*)pVertices,
g_pMeshSysMem->GetNumVertices(),
D3DXGetFVFVertexSize(g_pMeshSysMem->GetFVF()), &g_vObjectCenter,
&g_fObjectRadius);
   pVB->Unlock();
   SAFE_RELEASE(pVB);

   if(FAILED(hr))
      return hr;

   if(0 = = g_dwNumMaterials)
      return E_INVALIDARG;

   D3DXMatrixTranslation(&g_mCenterWorld, -g_vObjectCenter.x,
-g_vObjectCenter.y, -g_vObjectCenter.z);

   //Change the current directory to the.x's directory so
   //that the search can find the texture files.
   GetCurrentDirectory(MAX_PATH, wszWorkingDir);
   wszWorkingDir[MAX_PATH - 1] = L'\0';
   SetCurrentDirectory(wszMeshDir);

   //Get the array of materials out of the returned buffer, allocate a
   //texture array, and load the textures
   g_pMaterials = (D3DXMATERIAL*)g_pbufMaterials->GetBufferPointer();
   g_ppTextures = new LPDIRECT3DTEXTURE9[g_dwNumMaterials];

   for(UINT i = 0; i < g_dwNumMaterials; i++)
   {
      WCHAR strTexturePath[512] = L"";
      WCHAR* wszName;
      WCHAR wszBuf[MAX_PATH];
      wszName = wszBuf;
      MultiByteToWideChar(CP_ACP, 0, g_pMaterials[i].pTextureFilename, -1,
wszBuf, MAX_PATH);
      wszBuf[MAX_PATH - 1] = L'\0';
      DXUTFindDXSDKMediaFileCch(strTexturePath, 512, wszName);
      if(FAILED(D3DXCreateTextureFromFile(pd3dDevice, strTexturePath,
&g_ppTextures[i])))
         g_ppTextures[i] = NULL;
   }
   SetCurrentDirectory(wszWorkingDir);

   //Make sure there are normals, which are required for the
   //tessellation enhancement.
   if(!(g_pMeshSysMem->GetFVF() & D3DFVF_NORMAL))
   {
      ID3DXMesh* pTempMesh;
```

```
        V_RETURN(g_pMeshSysMem->CloneMeshFVF(g_pMeshSysMem->GetOptions(),
g_pMeshSysMem->GetFVF() | D3DFVF_NORMAL, pd3dDevice, &pTempMesh));
        D3DXComputeNormals(pTempMesh, NULL);

        SAFE_RELEASE(g_pMeshSysMem);
        g_pMeshSysMem = pTempMesh;
        }

        V_RETURN(GenerateEnhancedMesh(pd3dDevice, g_dwNumSegs));

        return S_OK;
}

bool SetTriangleCount(double k)
{
        unsigned int segmentCount = (int)sqrt((float)k/
(float)g_OriginalNumFaces);
        unsigned int targetFaceCount = 0;

        unsigned int faceCount = g_OriginalNumFaces * (segmentCount *
segmentCount);
        unsigned int faceCount2 = g_OriginalNumFaces * ((segmentCount + 1)
* (segmentCount + 1));

        if (abs((int)(faceCount - k)) < abs((int)(faceCount2 - k)))
        {
           targetFaceCount = faceCount;
        }
        else
        {
           targetFaceCount = faceCount2;
        }

        g_dwNumSegs = (int)sqrt((float)targetFaceCount/(float)g_
OriginalNumFaces);

        GenerateEnhancedMesh(g_pd3dDevice, g_dwNumSegs);

   WCHAR wszBuf[256];
   //StringCchPrintf(wszBuf, 256, L"Number of segments:%u", g_dwNumSegs);
   g_SampleUI.GetStatic(IDC_SEGMENTLABEL)->SetText(wszBuf);
     g_SampleUI.GetSlider(IDC_SEGMENT)->SetValue(g_dwNumSegs);

        return true;
}

bool SetShaderComplexity(double k)
{
        unsigned int shaderlevel = (unsigned int)(k);
        if (shaderlevel > 5)
           return false;
        g_CurrentTechnique = shaderlevel;
        g_SampleUI.GetComboBox(IDC_ACTIVEEFFECT)->SetSelectedByData((void*)
shaderlevel);
        return true;
}
```

```
void LoadCubeTexture(LPCWSTR FileName)
{
   WCHAR strPath[MAX_PATH];
   DXUTFindDXSDKMediaFileCch(strPath, MAX_PATH, FileName);

      IDirect3DCubeTexture9* cubeTexture;

   D3DXCreateCubeTextureFromFileEx(g_pd3dDevice, strPath, D3DX_DEFAULT,
1, 0, D3DFMT_A16B16G16R16F, D3DPOOL_MANAGED, D3DX_FILTER_NONE,
D3DX_FILTER_NONE, 0, NULL, NULL, &cubeTexture);

      SAFE_RELEASE(g_pCubeTexture);

      g_pCubeTexture = cubeTexture;
}

inline float GaussianDistribution(float x, float y, float rho)
{
   float g = 1.0f/sqrtf(2.0f * D3DX_PI * rho * rho);
   g * = expf(-(x * x + y * y)/(2 * rho * rho));

   return g;
}

//Auxiliary helper functions
inline int log2_ceiling(float val)
{
   int iMax = 256;
   int iMin = 0;

   while(iMax - iMin > 1)
   {
      int iMiddle = (iMax + iMin)/2;

      if(val > g_aPowsOfTwo[iMiddle])
         iMin = iMiddle;
      else
         iMax = iMiddle;
   }

   return iMax - 128;
}

inline VOID EncodeRGBE8(D3DXFLOAT16* pSrc, BYTE** ppDest)
{
   FLOAT r, g, b;

   r = (FLOAT)*(pSrc + 0);
   g = (FLOAT)*(pSrc + 1);
   b = (FLOAT)*(pSrc + 2);

   //Determine the largest colour component
   float maxComponent = max(max(r, g), b);
```

```
    //Round to the nearest integer exponent
    int nExp = log2_ceiling(maxComponent);

    //Divide the components by the shared exponent
    FLOAT fDivisor = (FLOAT)g_aPowsOfTwo[nExp + 128];

    r/= fDivisor;
    g/= fDivisor;
    b/= fDivisor;

    //Constrain the colour components
    r = max(0, min(1, r));
    g = max(0, min(1, g));
    b = max(0, min(1, b));

    //Store the shared exponent in the alpha channel
    D3DCOLOUR* pDestColour = (D3DCOLOUR*)*ppDest;
    *pDestColour = D3DCOLOUR_RGBA((BYTE)(r * 255), (BYTE)(g * 255), (BYTE)
(b * 255), nExp + 128);
    *ppDest + = sizeof(D3DCOLOUR);
}

//————————————————————————————————————————————————
inline VOID EncodeRGB16(D3DXFLOAT16* pSrc, BYTE** ppDest)
{
    FLOAT r, g, b;

    r = (FLOAT)*(pSrc + 0);
    g = (FLOAT)*(pSrc + 1);
    b = (FLOAT)*(pSrc + 2);

    //Divide the components by the multiplier
    r/= RGB16_MAX;
    g/= RGB16_MAX;
    b/= RGB16_MAX;

    //Constrain the colour components
    r = max(0, min(1, r));
    g = max(0, min(1, g));
    b = max(0, min(1, b));

    //Store
    USHORT* pDestColour = (USHORT*)*ppDest;
    *pDestColour++ = (USHORT)(r * 65535);
    *pDestColour++ = (USHORT)(g * 65535);
    *pDestColour++ = (USHORT)(b * 65535);

    *ppDest + = sizeof(UINT64);
}

HRESULT RetrieveTechHandles()
{
    DWORD dwNumTechniques = sizeof(TECH_HANDLES)/sizeof(D3DXHANDLE);

    CHAR strBuffer[MAX_PATH] = {0};

    D3DXHANDLE* pHandle = (D3DXHANDLE*)&g_aTechHandles;
```

```
   for(UINT t = 0; t < dwNumTechniques; t++)
   {
      StringCchPrintfA(strBuffer, MAX_PATH - 1, "%s", g_Techniques[t]);

      *pHandle++ = g_pEffect->GetTechniqueByName(strBuffer);
   }

   return S_OK;
}

//-----------------------------------------------------------------------
//Create any D3D9 resources that will live through a device reset
//(D3DPOOL_MANAGED) and aren't tied to the back buffer size
//-----------------------------------------------------------------------
HRESULT CALLBACK OnCreateDevice(IDirect3DDevice9* pd3dDevice, const
D3DSURFACE_DESC* pBackBufferSurfaceDesc, void* pUserContext)
{
   HRESULT hr;

   V_RETURN(g_DialogResourceManager.OnD3D9CreateDevice(pd3dDevice));
   V_RETURN(g_SettingsDlg.OnD3D9CreateDevice(pd3dDevice));

   g_pd3dDevice = pd3dDevice;

   D3DCAPS9 d3dCaps;
   pd3dDevice->GetDeviceCaps(&d3dCaps);
   if(!(d3dCaps.DevCaps & D3DDEVCAPS_NPATCHES))
   {
      //No hardware support. Disable the checkbox.
      g_bUseHWNPatches = false;
      g_SampleUI.GetCheckBox(IDC_HWNPATCHES)->SetChecked(false);
      g_SampleUI.GetCheckBox(IDC_HWNPATCHES)->SetEnabled(false);
   }
   else
      g_SampleUI.GetCheckBox(IDC_HWNPATCHES)->SetEnabled(true);

   //Initialize the font
   V_RETURN(D3DXCreateFont(pd3dDevice, 15, 0, FW_BOLD, 1, FALSE,
DEFAULT_CHARSET, OUT_DEFAULT_PRECIS, DEFAULT_QUALITY, DEFAULT_PITCH |
FF_DONTCARE, L"Arial", &g_pFont));
   g_DwShaderFlags = D3DXFX_NOT_CLONEABLE;
#if defined(DEBUG) || defined(_DEBUG)
      g_DwShaderFlags |= D3DXSHADER_DEBUG;
   #endif
#ifdef DEBUG_VS
      g_DwShaderFlags |= D3DXSHADER_FORCE_VS_SOFTWARE_NOOPT;
   #endif
#ifdef DEBUG_PS
      g_DwShaderFlags |= D3DXSHADER_FORCE_PS_SOFTWARE_NOOPT;
   #endif

      V_RETURN(LoadMesh(pd3dDevice, g_CurrentModel));
      V_RETURN(LoadEffect(pd3dDevice, L"EnhancedMesh.fx"));

   RetrieveTechHandles();
```

```
  //Determine which encoding modes this device can support
  IDirect3D9* pD3D = DXUTGetD3D9Object();
  DXUTDeviceSettings settings = DXUTGetDeviceSettings();

    LoadCubeTexture(g_CubeTextures[0]);
  V_RETURN(g_Skybox.OnCreateDevice(pd3dDevice, 50, g_pCubeTexture,
L"skybox.fx"));

  //Create the 1x1 white default texture
  V_RETURN(pd3dDevice->CreateTexture(1, 1, 1, 0, D3DFMT_A8R8G8B8,
D3DPOOL_MANAGED, &g_pDefaultTex, NULL));

  D3DLOCKED_RECT lr;
  V_RETURN(g_pDefaultTex->LockRect(0, &lr, NULL, 0));
  *(LPDWORD)lr.pBits = D3DCOLOUR_RGBA(255, 255, 255, 255);
  V_RETURN(g_pDefaultTex->UnlockRect(0));

  //Setup the camera's view parameters
  D3DXVECTOR3 vecEye(0.0f, 0.0f, -5.0f);
  D3DXVECTOR3 vecAt (0.0f, 0.0f, -0.0f);
  g_Camera.SetViewParams(&vecEye, &vecAt);

  return S_OK;
}

//----------------------------------------------------------------------
//Create any D3D9 resources that won't live through a device reset
//(D3DPOOL_DEFAULT) or that are tied to the back buffer size
//----------------------------------------------------------------------
HRESULT CALLBACK OnResetDevice(IDirect3DDevice9* pd3dDevice,
const D3DSURFACE_DESC* pBackBufferSurfaceDesc, void* pUserContext)
{
  HRESULT hr;
  int i = 0;

  V_RETURN(g_DialogResourceManager.OnD3D9ResetDevice());
  V_RETURN(g_SettingsDlg.OnD3D9ResetDevice());

  g_Skybox.OnResetDevice(pBackBufferSurfaceDesc);

  if(g_pFont)
    V_RETURN(g_pFont->OnResetDevice());
  if(g_pEffect)
    V_RETURN(g_pEffect->OnResetDevice());

  D3DFORMAT fmt = D3DFMT_UNKNOWN;
  switch(g_eEncodingMode)
  {
    case FP16:
      fmt = D3DFMT_A16B16G16R16F; break;
    case FP32:
      fmt = D3DFMT_A16B16G16R16F; break;
    case RGBE8:
      fmt = D3DFMT_A8R8G8B8; break;
    case RGB16:
      fmt = D3DFMT_A16B16G16R16; break;
  }
```

```
   hr = pd3dDevice->CreateTexture(pBackBufferSurfaceDesc->Width,
pBackBufferSurfaceDesc->Height, 1, D3DUSAGE_RENDERTARGET, fmt,
D3DPOOL_DEFAULT, &g_pTexRender, NULL);
   if(FAILED(hr))
      return hr;

   hr = pd3dDevice->CreateTexture(pBackBufferSurfaceDesc->Width/8,
pBackBufferSurfaceDesc->Height/8, 1, D3DUSAGE_RENDERTARGET,
D3DFMT_A8R8G8B8, D3DPOOL_DEFAULT, &g_pTexBrightPass, NULL);
   if(FAILED(hr))
      return hr;

   //Determine whether we can and should support a multisampling on the
HDR render target
   g_bUseMultiSample = false;
   IDirect3D9* pD3D = DXUTGetD3D9Object();
   if(!pD3D)
      return E_FAIL;

   DXUTDeviceSettings settings = DXUTGetDeviceSettings();

   g_bSupportsD16 = false;
   if(SUCCEEDED(pD3D->CheckDeviceFormat(settings.d3d9.AdapterOrdinal,
settings.d3d9.DeviceType, settings.d3d9.AdapterFormat,
D3DUSAGE_DEPTHSTENCIL, D3DRTYPE_SURFACE, D3DFMT_D16)))
   {
       if(SUCCEEDED(pD3D->CheckDepthStencilMatch(settings.d3d9.
AdapterOrdinal, settings.d3d9.DeviceType, settings.d3d9.AdapterFormat,
fmt, D3DFMT_D16)))
       {
          g_bSupportsD16 = true;
       }
   }
   g_bSupportsD32 = false;
   if(SUCCEEDED(pD3D->CheckDeviceFormat(settings.d3d9.AdapterOrdinal,
settings.d3d9.DeviceType, settings.d3d9.AdapterFormat,
D3DUSAGE_DEPTHSTENCIL, D3DRTYPE_SURFACE, D3DFMT_D32)))
   {
       if(SUCCEEDED(pD3D->CheckDepthStencilMatch(settings.d3d9.
AdapterOrdinal, settings.d3d9.DeviceType, settings.d3d9.AdapterFormat,
fmt, D3DFMT_D32)))
       {
          g_bSupportsD32 = true;
       }
   }
   g_bSupportsD24X8 = false;
   if(SUCCEEDED(pD3D->CheckDeviceFormat(settings.d3d9.AdapterOrdinal,
settings.d3d9.DeviceType, settings.d3d9.AdapterFormat,
D3DUSAGE_DEPTHSTENCIL, D3DRTYPE_SURFACE, D3DFMT_D24X8)))
   {
       if(SUCCEEDED(pD3D->CheckDepthStencilMatch(settings.d3d9.
AdapterOrdinal, settings.d3d9.DeviceType, settings.d3d9.AdapterFormat,
fmt, D3DFMT_D24X8)))
       {
          g_bSupportsD24X8 = true;
       }
   }
```

```
D3DFORMAT dfmt = D3DFMT_UNKNOWN;
if(g_bSupportsD16)
   dfmt = D3DFMT_D16;
else if(g_bSupportsD32)
   dfmt = D3DFMT_D32;
else if(g_bSupportsD24X8)
   dfmt = D3DFMT_D24X8;

if(dfmt ! = D3DFMT_UNKNOWN)
{
   D3DCAPS9 Caps;
   pd3dDevice->GetDeviceCaps(&Caps);

   g_MaxMultiSampleType = D3DMULTISAMPLE_NONE;
   for(D3DMULTISAMPLE_TYPE imst = D3DMULTISAMPLE_2_SAMPLES; imst < =
D3DMULTISAMPLE_16_SAMPLES;
      imst = (D3DMULTISAMPLE_TYPE)(imst + 1))
   {
      DWORD msQuality = 0;
      if(SUCCEEDED(pD3D->CheckDeviceMultiSampleType(Caps.AdapterOrdinal,
Caps.DeviceType, fmt, settings.d3d9.pp.Windowed, imst, &msQuality)))
      {
         g_bUseMultiSample = true;
         g_MaxMultiSampleType = imst;
         if(msQuality > 0)
            g_dwMultiSampleQuality = msQuality - 1;
         else
            g_dwMultiSampleQuality = msQuality;
      }
   }

   //Create the Multi-Sample floating point render target
   if(g_bUseMultiSample)
   {
      const D3DSURFACE_DESC* pBackBufferDesc =
DXUTGetD3D9BackBufferSurfaceDesc();
      hr = g_pd3dDevice->CreateRenderTarget(pBackBufferDesc->Width,
pBackBufferDesc->Height, fmt, g_MaxMultiSampleType,
g_dwMultiSampleQuality, FALSE, &g_pMSRT, NULL);
      if(FAILED(hr))
         g_bUseMultiSample = false;
      else
      {
         hr = g_pd3dDevice->CreateDepthStencilSurface(pBackBufferD
esc->Width, pBackBufferDesc->Height, dfmt, g_MaxMultiSampleType,
g_dwMultiSampleQuality, TRUE, &g_pMSDS, NULL);
         if(FAILED(hr))
         {
            g_bUseMultiSample = false;
            SAFE_RELEASE(g_pMSRT);
         }
      }
   }
}
```

```
   //For each scale stage, create a texture to hold the intermediate
   //results of the luminance calculation
   int nSampleLen = 1;
   for(i = 0; i < NUM_TONEMAP_TEXTURES; i++)
   {
      fmt = D3DFMT_UNKNOWN;
      switch(g_eEncodingMode)
      {
         case FP16:
            fmt = D3DFMT_R16F; break;
         case FP32:
            fmt = D3DFMT_R32F; break;
         case RGBE8:
            fmt = D3DFMT_A8R8G8B8; break;
         case RGB16:
            fmt = D3DFMT_A16B16G16R16; break;
      }

      hr = pd3dDevice->CreateTexture(nSampleLen, nSampleLen, 1,
D3DUSAGE_RENDERTARGET, fmt, D3DPOOL_DEFAULT, &g_apTexToneMap[i], NULL);
      if(FAILED(hr))
         return hr;

      nSampleLen * = 3;
   }

   //Create the temporary blooming effect textures
   for(i = 0; i < NUM_BLOOM_TEXTURES; i++)
   {
      hr = pd3dDevice->CreateTexture(pBackBufferSurfaceDesc->Width/8,
pBackBufferSurfaceDesc->Height/8, 1, D3DUSAGE_RENDERTARGET,
D3DFMT_A8R8G8B8, D3DPOOL_DEFAULT, &g_apTexBloom[i], NULL);
      if(FAILED(hr))
         return hr;
   }

   //Create a sprite to help batch calls when drawing many lines of text
   V_RETURN(D3DXCreateSprite(pd3dDevice, &g_pTextSprite));

   V_RETURN(GenerateEnhancedMesh(pd3dDevice, g_dwNumSegs));

   if(g_bWireframe)
      pd3dDevice->SetRenderState(D3DRS_FILLMODE, D3DFILL_WIREFRAME);
   else
      pd3dDevice->SetRenderState(D3DRS_FILLMODE, D3DFILL_SOLID);

   //Setup the camera's projection parameters
   float fAspectRatio = pBackBufferSurfaceDesc->Width/(FLOAT)
pBackBufferSurfaceDesc->Height;
   g_Camera.SetProjParams(D3DX_PI/4, fAspectRatio, 0.1f, 1000.0f);
   g_Camera.SetWindow(pBackBufferSurfaceDesc->Width,
pBackBufferSurfaceDesc->Height);

   g_HUD.SetLocation(pBackBufferSurfaceDesc->Width - 170, 0);
   g_HUD.SetSize(170, 170);
```

```
   g_SampleUI.SetLocation(pBackBufferSurfaceDesc->Width - 170,
pBackBufferSurfaceDesc->Height - 350);
   g_SampleUI.SetSize(170, 300);

   return S_OK;
}

//─────────────────────────────────────────────────────────────
//Handle updates to the scene. This is called regardless of which
//D3D API is used
//─────────────────────────────────────────────────────────────
void CALLBACK OnFrameMove(double fTime, float fElapsedTime,
void* pUserContext)
{
   IDirect3DDevice9* pd3dDevice = DXUTGetD3D9Device();

   //Update the camera's position based on user input
   g_Camera.FrameMove(fElapsedTime);

   pd3dDevice->SetTransform(D3DTS_WORLD, g_Camera.GetWorldMatrix());
   pd3dDevice->SetTransform(D3DTS_VIEW, g_Camera.GetViewMatrix());

   g_pEffect->SetValue("g_vEyePt", g_Camera.GetEyePt(),
sizeof(D3DXVECTOR3));
}

//─────────────────────────────────────────────────────────────
//Render the scene using the D3D9 device
//─────────────────────────────────────────────────────────────
void CALLBACK OnFrameRender(IDirect3DDevice9* pd3dDevice, double
fTime, float fElapsedTime, void* pUserContext)
{
   //If the settings dialog is being shown, then
   //render it instead of rendering the app's scene
   if(g_SettingsDlg.IsActive())
   {
      g_SettingsDlg.OnRender(fElapsedTime);
      return;
   }

   HRESULT hr;
   D3DXMATRIXA16 mWorld;
   D3DXMATRIXA16 mWorldI;
   D3DXMATRIXA16 mView;
   D3DXMATRIXA16 mProj;
   D3DXMATRIXA16 mWorldViewProjection;

   //Clear the render target and the zbuffer
   V(pd3dDevice->Clear(0, NULL, D3DCLEAR_TARGET | D3DCLEAR_ZBUFFER,
D3DCOLOUR_ARGB(0, 0, 0, 0), 1.0f, 0));

   //Render the scene
   if(SUCCEEDED(pd3dDevice->BeginScene()))
   {
#ifdef NETWORK_CONTROL
         //Control actions
```

```
              SetTriangleCount(dr.data[0]);
              SetShaderComplexity(dr.data[1]);
#endif
          //Get the projection & view matrix from the camera class
      mWorld = *g_Camera.GetWorldMatrix();
      mProj = *g_Camera.GetProjMatrix();
      mView = *g_Camera.GetViewMatrix();

      mWorldViewProjection = g_mCenterWorld * mWorld * mView * mProj;

      g_Skybox.Render(&mWorldViewProjection);

   //Update the effect's variables. Instead of using strings, it would
   //be more efficient to cache a handle to the parameter by calling
   //ID3DXEffect::GetParameterByName
      V(g_pEffect->SetMatrix("g_mWorldViewProjection",
&mWorldViewProjection));
      V(g_pEffect->SetMatrix("g_mWorld", &mWorld));
      V(g_pEffect->SetMatrix("g_mWorldI", &mWorldI));
      V(g_pEffect->SetMatrix("g_mView", &mView));
      V(g_pEffect->SetMatrix("g_mProj", &mProj));
         V(g_pEffect->SetFloat("g_fTime", (float)fTime));
         V(g_pEffect->SetFloat("g_fModelReflectivity",
g_fModelReflectivity));

      if(g_bUseHWNPatches)
      {
          float fNumSegs;

          fNumSegs = (float)g_dwNumSegs;
          pd3dDevice->SetNPatchMode(fNumSegs);
      }

      UINT cPasses;

          switch (g_CurrentTechnique)
          {
          case 0:
            g_pEffect->SetTechnique(g_pCurTechnique->XRay);
            break;
          case 1:
            g_pEffect->SetTechnique(g_pCurTechnique->SimpleLighting);
            break;
          case 2:
            g_pEffect->SetTechnique(g_pCurTechnique->SpecularLighting);
            break;
          case 3:
            g_pEffect->SetTechnique(g_pCurTechnique->ToonEffect);
            break;
          case 4:
            g_pEffect->SetTechnique(g_pCurTechnique->Reflect);
            break;
          case 5:
            g_pEffect->SetTechnique(g_pCurTechnique->ReflectSpecular);
            break;
          }
        g_pEffect->SetTexture("g_tCube", g_Skybox.GetEnvironmentMap());
```

```
      V(g_pEffect->Begin(&cPasses, 0));
      for(UINT p = 0; p < cPasses; ++p)
      {
         V(g_pEffect->BeginPass(p));

         //set and draw each of the materials in the mesh
         for(UINT i = 0; i < g_dwNumMaterials; i++)
         {
            V(g_pEffect->SetVector("g_vDiffuse",
(D3DXVECTOR4*)&g_pMaterials[i].MatD3D.Diffuse));
            if(g_ppTextures[i])
            {
               V(g_pEffect->SetTexture("g_txScene", g_ppTextures[i]));
            }
            else
            {
               V(g_pEffect->SetTexture("g_txScene", g_pDefaultTex));
            }
               V(g_pEffect->CommitChanges());
            g_pMeshEnhanced->DrawSubset(i);
         }

         V(g_pEffect->EndPass());
      }
      V(g_pEffect->End());

      if(g_bUseHWNPatches)
      {
         pd3dDevice->SetNPatchMode(0);
      }

      RenderText();
      V(g_HUD.OnRender(fElapsedTime));
      V(g_SampleUI.OnRender(fElapsedTime));

      V(pd3dDevice->EndScene());
   }

#ifdef NETWORK_CONTROL
      ds.data[0] = DXUTGetFPS();  //Sending the frame rate to the
Controller
      //printf("My FPS is:%f\n", ds.data[0]);
#endif
}

//————————————————————————————————————————————————————
//Render the help and statistics text. This function uses the
//ID3DXFont interface for efficient text rendering.
//————————————————————————————————————————————————————
void RenderText()
{
   //The helper object simply helps keep track of text position,
   //and colour and then it calls pFont->DrawText(m_pSprite, strMsg, -1,
   //&rc, DT_NOCLIP, m_clr);
   //If NULL is passed in as the sprite object, then it will work
   //however the pFont->DrawText() will not be batched together.
   //Batching calls will improves performance.
```

```
CDXUTTextHelper txtHelper(g_pFont, g_pTextSprite, 15);

//Output statistics
txtHelper.Begin();
txtHelper.SetInsertionPos(5, 5);
txtHelper.SetForegroundColour(D3DXCOLOUR(1.0f, 1.0f, 0.0f, 1.0f));
txtHelper.DrawTextLine(DXUTGetFrameStats(DXUTIsVsyncEnabled()));
txtHelper.DrawTextLine(DXUTGetDeviceStats());

//Draw help
if(g_bShowHelp)
{
    const D3DSURFACE_DESC* pd3dsdBackBuffer =
DXUTGetD3D9BackBufferSurfaceDesc();
    txtHelper.SetInsertionPos(10, pd3dsdBackBuffer->Height - 15 * 9);
    txtHelper.SetForegroundColour(D3DXCOLOUR(1.0f, 0.75f, 0.0f, 1.0f));
    txtHelper.DrawTextLine(L"Controls (F1 to hide):");

    txtHelper.SetInsertionPos(40, pd3dsdBackBuffer->Height - 15 * 8);

        txtHelper.DrawTextLine(L"F4: Load next mesh\n"
                               L"1,2,3,4,5,6: Load mesh\n"
                               L"Rotate mesh: Left click drag\n"
                               L"Rotate camera: right click drag\n"
                               L"Zoom: Mouse wheel\n"
                               L"Quit: ESC");
}
else
{
    txtHelper.SetForegroundColour(D3DXCOLOUR(1.0f, 1.0f, 1.0f, 1.0f));
    txtHelper.DrawTextLine(L"Press F1 for help");
}

    float fps = DXUTGetFPS();
    //Write to logfile
    //logfile << fps << " " << g_pMeshEnhanced->GetNumVertices() <<
std::endl; //SISO
    logfile << fps << " " << g_pMeshEnhanced->GetNumVertices()/*<< "
" << g_CurrentTechnique*/<< std::endl; //MISO/MIMO

txtHelper.SetForegroundColour(D3DXCOLOUR(1.0f, 0.75f, 0.0f, 1.0f));
txtHelper.SetInsertionPos(10, 65);
txtHelper.DrawFormattedTextLine(L"NumSegs:%d\n", g_dwNumSegs);
txtHelper.DrawFormattedTextLine(L"NumFaces:%d\n",
(g_pMeshEnhanced == NULL) ? 0 : g_pMeshEnhanced->GetNumFaces());
txtHelper.DrawFormattedTextLine(L"NumVertices:%d\n",
(g_pMeshEnhanced == NULL) ? 0 : g_pMeshEnhanced->GetNumVertices());
    txtHelper.DrawFormattedTextLine(L"FPS:%f\n", fps);

txtHelper.End();
}

//─────────────────────────────────────────────────────────────
//Handle messages to the application
//─────────────────────────────────────────────────────────────
LRESULT CALLBACK MsgProc(HWND hWnd, UINT uMsg, WPARAM wParam, LPARAM
lParam, bool* pbNoFurtherProcessing, void* pUserContext)
```

```
{
  //Always allow dialog resource manager calls to handle global
  //messages so GUI state is updated correctly
  *pbNoFurtherProcessing = g_DialogResourceManager.MsgProc(hWnd, uMsg,
wParam, lParam);
  if(*pbNoFurtherProcessing)
    return 0;

  if(g_SettingsDlg.IsActive())
  {
    g_SettingsDlg.MsgProc(hWnd, uMsg, wParam, lParam);
    return 0;
  }

  //Give the dialogs a chance to handle the message first
  *pbNoFurtherProcessing = g_HUD.MsgProc(hWnd, uMsg, wParam, lParam);
  if(*pbNoFurtherProcessing)
    return 0;
  *pbNoFurtherProcessing = g_SampleUI.MsgProc(hWnd, uMsg, wParam,
lParam);
  if(*pbNoFurtherProcessing)
    return 0;

  //Pass all remaining windows messages to camera so it can respond to
user input
  g_Camera.HandleMessages(hWnd, uMsg, wParam, lParam);

  return 0;
}

//---------------------------------------------------------------
//Handle key presses
//---------------------------------------------------------------
void CALLBACK KeyboardProc(UINT nChar, bool bKeyDown, bool bAltDown,
void* pUserContext)
{
    IDirect3DCubeTexture9* pCubeTexture = NULL;
  if(bKeyDown)
  {
    switch(nChar)
    {
      case VK_F1:
        g_bShowHelp = !g_bShowHelp; break;
          case VK_F4:
            if (++g_CurrentModel = = g_ModelCount)
              g_CurrentModel = 0;
            LoadMesh(DXUTGetD3D9Device(), g_CurrentModel);
            break;
          case '1':
            g_CurrentModel = 0;
            LoadMesh(DXUTGetD3D9Device(), g_CurrentModel);
            break;
          case '2':
            g_CurrentModel = 1;
            LoadMesh(DXUTGetD3D9Device(), g_CurrentModel);
            break;
```

```
                    case '3':
                      g_CurrentModel = 2;
                      LoadMesh(DXUTGetD3D9Device(), g_CurrentModel);
                      break;
                    case '4':
                      g_CurrentModel = 3;
                      LoadMesh(DXUTGetD3D9Device(), g_CurrentModel);
                      break;
                    case '5':
                      g_CurrentModel = 4;
                      LoadMesh(DXUTGetD3D9Device(), g_CurrentModel);
                      break;
                    case '6':
                      g_CurrentModel = 5;
                      LoadMesh(DXUTGetD3D9Device(), g_CurrentModel);
                      break;
                    case 'B':
                      WCHAR strPath[MAX_PATH];

                      DXUTFindDXSDKMediaFileCch(strPath, MAX_PATH, L"Light
Probes\\uffizi_cross.dds");
                      D3DXCreateCubeTextureFromFileEx(g_pd3dDevice, strPath,
D3DX_DEFAULT, 1, 0, D3DFMT_A16B16G16R16F, D3DPOOL_MANAGED,
D3DX_FILTER_NONE, D3DX_FILTER_NONE, 0, NULL,
                          NULL, &pCubeTexture);
                      g_Skybox.SetEnvironmentMap(pCubeTexture);
                      break;
                    case 'N':
                      SetShaderComplexity(4);
                      break;
            }
        }
}

//————————————————————————————————————————————————————
//Handles the GUI events
//————————————————————————————————————————————————————
void CALLBACK OnGUIEvent(UINT nEvent, int nControlID,
CDXUTControl* pControl, void* pUserContext)
{
    int index;

    switch(nControlID)
    {
      case IDC_TOGGLEFULLSCREEN:
        DXUTToggleFullScreen(); break;
      case IDC_TOGGLEREF:
        DXUTToggleREF(); break;
      case IDC_CHANGEDEVICE:
        g_SettingsDlg.SetActive(!g_SettingsDlg.IsActive()); break;
      case IDC_FILLMODE:
      {
        g_bWireframe = ((CDXUTComboBox*)pControl)->GetSelectedData()
! = 0;
```

```
        IDirect3DDevice9* pd3dDevice = DXUTGetD3D9Device();
        pd3dDevice->SetRenderState(D3DRS_FILLMODE, g_bWireframe ?
D3DFILL_WIREFRAME : D3DFILL_SOLID);
        break;
    }
    case IDC_SEGMENT:
        g_dwNumSegs = ((CDXUTSlider*)pControl)->GetValue();
        WCHAR wszBuf[256];
        StringCchPrintf(wszBuf, 256, L"Number of segments:%u",
g_dwNumSegs);
        g_SampleUI.GetStatic(IDC_SEGMENTLABEL)->SetText(wszBuf);
        GenerateEnhancedMesh(DXUTGetD3D9Device(), g_dwNumSegs);
        break;
            case IDC_HWNPATCHES:
                g_bUseHWNPatches = ((CDXUTCheckBox*)pControl)-
>GetChecked();
                GenerateEnhancedMesh(DXUTGetD3D9Device(), g_dwNumSegs);
                break;
            case IDC_CUBETEXTURE:
                index = (int)((CDXUTComboBox*)pControl)->GetSelectedData();
                LoadCubeTexture(g_CubeTextures[index - 1]);
                g_Skybox.SetEnvironmentMap(g_pCubeTexture);
                break;
            case IDC_MODELREFLECTIVITY:
                g_fModelReflectivity = (float)((CDXUTSlider*)pControl)-
>GetValue()/100.0f;
                StringCchPrintf(wszBuf, 256, L"Model Reflectivity:%u",
((CDXUTSlider*)pControl)->GetValue());
                g_SampleUI.GetStatic(IDC_MODELREFLECTIVITYLABEL)-
>SetText(wszBuf);
                break;
            case IDC_ACTIVEEFFECT:
                index = (int)((CDXUTComboBox*)pControl)->GetSelectedData();
        g_CurrentTechnique = index;

        break;
    }
}


//------------------------------------------------------------
//Release D3D9 resources created in the OnResetDevice callback
//------------------------------------------------------------
void CALLBACK OnLostDevice(void* pUserContext)
{
  g_DialogResourceManager.OnD3D9LostDevice();
  g_SettingsDlg.OnD3D9LostDevice();

  g_Skybox.OnLostDevice();

  if(g_pFont)
    g_pFont->OnLostDevice();
  if(g_pEffect)
    g_pEffect->OnLostDevice();
```

```
        SAFE_RELEASE(g_pTextSprite);
        SAFE_RELEASE(g_pMeshEnhanced);

        SAFE_RELEASE(g_pMSRT);
        SAFE_RELEASE(g_pMSDS);

        SAFE_RELEASE(g_pTexRender);
        SAFE_RELEASE(g_pTexBrightPass);

            int i = 0;

        for(i = 0; i < NUM_TONEMAP_TEXTURES; i++)
        {
            SAFE_RELEASE(g_apTexToneMap[i]);
        }

        for(i = 0; i < NUM_BLOOM_TEXTURES; i++)
        {
            SAFE_RELEASE(g_apTexBloom[i]);
        }
}

//——————————————————————————————————————————————————————————————
//Release D3D9 resources created in the OnCreateDevice callback
//——————————————————————————————————————————————————————————————
void CALLBACK OnDestroyDevice(void* pUserContext)
{
    g_DialogResourceManager.OnD3D9DestroyDevice();
    g_SettingsDlg.OnD3D9DestroyDevice();
    SAFE_RELEASE(g_pEffect);
    SAFE_RELEASE(g_pFont);

    for(UINT i = 0; i < g_dwNumMaterials; i++)
        SAFE_RELEASE(g_ppTextures[i]);

    SAFE_RELEASE(g_pDefaultTex);
    SAFE_DELETE_ARRAY(g_ppTextures);
    SAFE_RELEASE(g_pMeshSysMem);
    SAFE_RELEASE(g_pbufMaterials);
    SAFE_RELEASE(g_pbufAdjacency);
    g_dwNumMaterials = 0L;

    g_Skybox.OnDestroyDevice();
}
```

Annex B: Patent for Application Method and System for Adaptive Control of Real-Time Computer Graphics Rendering[*]

TITLE OF INVENTION

Method and System for Adaptive Control of Real-Time Computer Graphics Rendering

FIELD OF INVENTION

This invention provides a method for controlling real-time computer generated graphics (known as rendering) such that it is able to achieve user-defined performance objectives without the user's intervention.

BACKGROUND OF INVENTION

Computer generated graphics is required in many interactive digital media applications today. The requirement for this process to be "real-time" is based upon the need for adequate response to the user in a continuous feedback loop. One key performance metric in interactive computer graphics systems is the time taken to render one frame of the animation sequence. Current techniques in interactive rendering do not guarantee consistent resource and time control in this respect. This is inhibitive in many aspects of computer software related to real-time computer graphics rendering.

In the present invention, a modelling framework is described for real-time computer graphics rendering. This modelling framework forms the first part of the entire method to control real-time computer graphics rendering. A control system is described in the present context which consists of the aforementioned rendering process model (also known as the "plant") and a controller module. Based on this method, the control system is capable of controlling the rendering process whether it is implemented in the local computer hardware or a remote computer device via a communication channel.

---

* U.S. Patent Office provisional application dated 13 July 2011, US61/507,486.

SUMMARY OF INVENTION

The object of the present invention is to provide a method to control the rendering of computer-generated graphics in real time for interactive applications.

In the present context, the rendering process refers to a set of computer program routines that support the generation of a sequence of images such that they create the impression of an animation. The rendering process is by convention a piece of software that operates on input in the form of data structures which describe the geometry of an object in 3D space and the quality of its visual appearance.

According to Figure B.1 the input to the rendering process may consist of one element or a set of elements as long as each element is independent of the others and each element exerts an observable influence on the output of the rendering process. The output of the rendering process may consist of one element or a set of elements. Both the input and output to the rendering process are measurable quantities. The input to the rendering process is a variable or a set of variables that can be changed during the execution of the application. The disturbance described in Figure B.1 refers to any auxiliary process that runs in the same environment as the rendering process. For example, the operating system's kernel processes run in the background which is mandatory for the computer device to function normally.

In the context of Figure B.1, the rendering process model can be expressed as, but is not limited to, a polynomial equation or a state space representation through an iterative process that involves regressive computation of the rendering process's previous input and outputs. The user defines the actual process input variables and output quantities based on the desired performance objectives and the controller design. More importantly, all input and output to the system must be measurable quantities.

In Figure B.2, the controller 201 is a module that may be implemented as a piece of software or a computer program that executes a set of computer instructions which is built into an embedded computer subsystem. The purpose of the controller is to adjust the input to the plant (the rendering process) such that the output of the plant 202 can be driven to meet a certain performance objective. The controller receives an input which is the difference between the user defined reference 203 and the current output from the plant. The controller is typically designed with saturation limits to prevent the system from swinging beyond normal operating range. The controller design is not limited to any particular control algorithm or a combination of such algorithms as long as the purpose of the controller is achieved. In the same spirit, the controller and the plant implementation is not limited to any specific programming language or software toolkit.

In Figure B.3 a deployment of the control system in a computer device is shown with the key components as the shared memory 301, the controller 302 and the plant 303. The controller sends its control action to the rendering process (the plant) via the shared memory where this value will thereafter be copied into the execution space of the rendering process. Similarly, the rendering process will write the values from its output into the same shared memory area where the controller will access, to copy these values for the error computation. In the context of this invention, the data access method is not limited to shared memory but variants of common memory access methods provided by any operating system such as pipes or any inter-process communication technique.

In Figure B.4, the control system is deployed in a distributed computing environment where the controller 401 and the plant 402 (the rendering process) are executed in different physical machine locations. In the context of this invention, the deployment platform is not limited to any particular operating system or 3D rendering toolkit. The controller and plant communication is realised through the external network infrastructure 403 that may be instituted with wired or wireless connection capability. The communication link 404 between the controller and the plant is driven by software routines using suitable protocol-based transmission such as and not limited to TCP and other IP-based standards. The software implementation supporting such a communication method can be of the client-server or peer-to-peer or any other architecture as long as the objective for reliable data transmission is supported.

BRIEF DESCRIPTIONS OF FIGURES

The present invention will now be further described with reference to the figures, wherein:

Figure B.1 illustrates the open-loop system model of the rendering process with the input and output of the system and the inherent disturbance arising from other processes that may be running in the computing environment.

Figure B.2 illustrates the closed-loop control system with feedback. The controller is introduced to ensure that the error between the output and the performance objective is eventually removed.

Figure B.3 illustrates the deployment of this control system in a single computer device. Both the controller and plant are software processes that run in the common/shared memory address space and communication between the controller and plant is done via shared memory.

Figure B.4 illustrates the deployment of this control system in a remote/distributed setting. In this scenario, the controller and the plant are running in separate and different computer machines. Communication between the controller and the plant is done via the network infrastructure which links the two computers.

DETAILED DESCRIPTIONS OF FIGURES

The present invention provides a method for automatic control of the real-time computer graphics rendering process such that it is able to consistently meet a certain performance objective. This is particularly important in many interactive applications where user's input to the application is processed and the response (output of the rendering process) is sent back to the user promptly. In cases where the rendering process takes unduly long time, the generated animation sequence will look "laggy" and thereby affect the user's visual and usage experience of the application.

Figure B.1 illustrates the fundamental system concept of the rendering process 101. Each rendering process can receive an input vector 102 and generates an output vector 103. A vector may consist of one or more elements. Since the rendering process is basically run on a computer device, there may be other processes that share the resources on the computer. The effect on the rendering process attributed by these external processes is defined as the disturbance 104 to the system.

Figure B.2 illustrates schematically the fundamental control system in a closed-loop configuration. The controller module 201 works on the error between the output 204 of the plant 202 and the reference 203 (performance objective). Depending on the design, the computed output of the controller module 201 will be fed into the plant 202 such that the plant's output may be regulated to the reference 203. This process is iterative until the error between the plant's output and the user-defined reference diminishes to a negligible value.

Figure B.3 illustrates the control system in componentised form localised within a computer device. The controller 301 and the plant 302 share the resources from this computer device, such as memory, data bus, and main processor's computation bandwidth. The controller and plant are connected for data exchange via the main memory using the shared memory 301 within the same execution space.

Figure B.4 illustrates the framework by which the control system is deployed in a distributed computing environment. The controller 401 resides in a different computer device from the plant 402 (the rendering process). The controller and the plant are linked via an external network 403. The control action and the plant's output are routed via bidirectional digital channel data 404 over this network.

CONTROL DESIGN AND MECHANISM

Due to the complexity in modern computer graphics hardware, rendering processes may not exhibit linear properties over certain operating ranges. The present invention describes a design technique that yields a controller which is capable of handling such non-linearity during the system's operation. The approach consists of two strategies:

    I. PID gain scheduling
    II. Fuzzy control

The design process commences with collection of a qualified set of input–output data pairs. The qualifications of the input and output variables are contingent upon whether the quantities are both measureable and controllable. The data generation process involves selecting a range of inputs that are sufficient to drive the dynamics of the rendering system. The derivation of the system model is based on the system identification methodology where the model may be represented in a linear auto-regressive (ARX) model or its corresponding state space representation.

I. PID GAIN SCHEDULING

After collecting the steady-state values of the input–output data, they are plotted against each other as shown in Figure B.5. The example shows the output (frame rate) is plotted against the input (vertex count). Empirically, the input–output relationship is typically non-linear. The gain scheduling technique proposed in this invention requires piece-wise approximation of non-linear curves using straight line segments. Each segment represents a linear region of operation by which linear time-invariant dynamic models may be derived using the aforementioned system identification technique.

To obtain the individual line segments for curves, we can describe this non-linear relationship represented by a polynomial model:

$$y = \sum_{i=1}^{n+1} p_i x^{n+1-i}, \quad u_0 \leq x \leq u_N \tag{1}$$

where n is the degree of the polynomial and $(n + 1)$ is the degree that gives the highest power of the predictor variable. Since straight line segments are used to fit the curve, the order of the polynomial is chosen as 1. The objective is to derive a series of line segments which fulfills the approximation of this relationship by the following:

$$y = \begin{cases} a_1 + b_1 x & u_0 \leq x \leq u_1 \\ a_2 + b_2 x & u_1 \leq x \leq u_2 \\ \cdots & \cdots \\ a_N + b_N x & u_{N-1} \leq x \leq u_N \end{cases} \tag{2}$$

where the variables a and b are to be found that minimise the following equation (a constrained optimisation problem):

$$F\left(a_1, a_2, \dots, a_N, b_1, b_2, \dots, b_N, u_1, u_2, \dots, u_{N-1}\right)$$

$$= \sum_{j=1}^{N} \int_{u_{j-1}}^{u_j} \left(f(x) - a_j - b_j x\right)^2 dx \tag{3}$$

and the right hand side of the equation represents the least square error of the approximation.

Given the solution to the optimisation problem in Equation (3), the input–output data pairs in each line segment shall be used for the derivation of the corresponding system model which may be expressed as in the following state space representation or its ARX model representation as shown in Equations (4), (5), and (6), respectively:

$$x(k+1) = Ax(k) + Bu(k) \tag{4}$$

$$y(k) = Cx(k) + Du(k) \tag{5}$$

Here x is the state variable of the system, u is the input to the system, y is the output of the system, and k is the time step. The ARX model representation is given by

$$y(t) + a_1 y(t-1) + \dots a_{n_a} y(t - n_a) = b_1 u(t - n_k) + \dots + b_{n_b} u(t - n_k - n_b + 1) \tag{6}$$

where:

$a_1 \ldots a_{n_a}$ and $b_1 \ldots b_{n_b}$ are parameters to be estimated.

$y(t)$ is the output of the system at time t.

$y(t-1) \ldots y(t-n_a)$ are the previous outputs on which the current output depends.

$u(t-n_k) \ldots u(t-n_k-n_b+1)$ are the previous inputs on which the current output depends.

n_a is the number of poles or the order of the system.

n_b is the number of zeroes plus one.

n_k is the delay in the system.

The proportional, integral, and derivative (PID) controller is well reputed for its adoption in over 90% of the world's real control systems. There are several advantages in using the PID controller, namely its efficiency attributed by the relatively simpler computation and the ease of implementation compared to other more elaborate control schemes. In brief, the PID control action in a closed-loop feedback system takes the form (parallel mode):

$$u(t) = K_p e(t) + K_i \int_0^t e(\tau)d\tau + K_d \frac{d}{dt}e(t) \tag{7}$$

where $K_i = \dfrac{K_p}{T_i}$ and $K_d = K_p T_d$ with T_i and T_d as the time constants of the integral and derivative controls. At the implementation level, the PID controller's discrete time form may be expressed as:

$$u(n) = K_p e(n) + \frac{K_p T}{T_i} \sum_{k=0}^{n} e(k) + \frac{K_p T_d}{T}\left(e(n)-e(n-1)\right) \tag{8}$$

where T is the sampling period and

$$K_i = \frac{K_p T}{T_i}, K_d = \frac{K_p T_d}{T}$$

where $u(n)$ is the control action. The PID controller's gain values may be derived either empirically via trial and adjustments or by using the model derived in the previous section in a closed-loop feedback system as shown in Figure B.2 with an auto-tuning algorithm.

The derivation of both the system model and the PID controller is exercised for each linear operation range corresponding to the line segments derived from the solution to Equation (3). By cascading the series of PID controllers, an overall control system may be derived as shown in Figure B.6. The object 601 represents the cascaded PID controller array in which only one PID controller is active at any time. The object 602 represents a switch agent that channels and activates the appropriate PID controller based on the operating point of the rendering system. Mechanisms may also be incorporated in to achieve the so-called bump-less transfer that smooths the abrupt changes in the behaviour of the system when switching among the controllers occurs.

II. FUZZY CONTROL (MODEL-LESS CONTROL)

The primary benefit offered by the fuzzy control paradigm is its ability to emulate human control based on linguistic variables and a set of intuitive expert rules used as the decision or inference system. In comparison to conventional control techniques, the advantages of the fuzzy control paradigm are twofold. First, there is no requirement for a mathematical model of the system to be controlled. This is especially important and useful as it may be difficult to derive certain process models due to their complex dynamics and when some systems cannot be modelled using first principles. Second, the fuzzy controller itself works on relatively straightforward computation and it can be developed to handle non-linear processes empirically in practice without the need for complicated mathematics. These advantages translate to its appeal as a practical solution to real world control problems in terms of implementation.

The development of fuzzy control system begins with the two key components: the input–output membership functions describing the properties of the system (fuzzy sets) based on linguistic variables and the rule base which relates the input–output sets. Given an antecedent and consequent relationship between an input y to a SISO system's output u using linguistic descriptions of their properties, this may be represented as:

$$IF \ y \in Y_j \ THEN \ u \in U_j \tag{9}$$

In each universe of discourse U_i and Y_i, u_i and y_i exist taking on values with corresponding linguistic variables $\widetilde{u}_i$ and $\widetilde{y}_i$ which describe the characteristics of the variables. Suppose $\widetilde{A}_i^j$ denotes the jth linguistic value of the linguistic variable $\widetilde{u}_i$ defined over the universe of discourse U_i. If the assumption that there exist many linguistic values defined in U_i, then the linguistic variable $\widetilde{u}_i$ which takes on the elements from the set of linguistic values may be denoted by Equation (9).

$$\widetilde{A}_i = \left\{ \widetilde{A}_i^j : j = 1, 2, ..., N_i \right\} \tag{10}$$

In the same manner, we can consider $\widetilde{B}_i^j$ to denote the jth value of the linguistic variable $\widetilde{y}_i$ defined over the universe of discourse Y_i. $\widetilde{y}_i$ may be represented by elements taken from the set of linguistic values denoted by the following equation:

$$\widetilde{B}_i = \left\{ \widetilde{B}_i^p : p = 1, 2, ..., M_i \right\} \tag{11}$$

Given a condition where all the premise terms are used in every rule and a rule is formed for each possible combination of premise elements, then we have rule set with N_i number of rules that can be expressed as:

$$\prod_{i=1}^{n} N_i = N_1 \cdot N_2 \cdot ... \cdot N_n \qquad (12)$$

Given the membership functions, the conversion of a crisp input value into its corresponding fuzzy value is known as fuzzification. The defuzzification of the resultant fuzzy set from the inference system to a quantifiable value may be done using the centroid (centre of gravity) method. The principle is to select the value in the resultant fuzzy set such that it would lead to the smallest error on average given any criterion. To determine y^* the least square method can be used and the square of the error is accompanied by the weightage of the grade of the membership $\mu_B(u)$. Therefore, the defuzzified output, y^* may be obtained by finding the solution to the following equation.

$$y^* = \arg \min_{y^*} \int_U \mu_B(y)(y^* - y)^2 du \qquad (13)$$

Differentiating with respect to y^* and equating the derivative to zero yields

$$y^* = \frac{\int_Y y\mu_B(y)dy}{\int_Y \mu_B(y)dy} \qquad (14)$$

which gives the value of the abscissa of the centre of gravity of the area below the membership function $\mu_B(u)$.

The derivation of the membership functions is based on intuitive recognition of the fundamental relationship between input and output of the rendering system. In the context of the present invention, for example, there is an inverse relationship between the frame rate and the total number of vertices used in the rendering process.

Figure B.7 indicates how this relationship may be developed in the form of a combination of sigmoid functions for both the input and output variables. The diagram 701 illustrates the membership functions used for the input variable. There are two function curves used for the linguistic value of the FPS error input variable. The function curve at the left is to describe the extent of *high* and the one at the right is used to describe the extent of *low*. In a similar manner, the diagram 702 shows the membership functions for the output variable, which is the vertex count. The rule base of the fuzzy inference rule set relating the input and output membership functions is shown in object 703.

In the same spirit as the closed-loop control feedback system shown in Figure B.2, a fuzzy controller-based rendering system may be constructed using the aforementioned approach and using the derived fuzzy controller as the controller block in Figure B.2.

CLAIMS (PRELIMINARY)

1. A method that defines correlation between the single or multiple inputs and single or multiple outputs of the computer graphics rendering process via a system's approach whereby:
    a. The inputs and outputs of the rendering process can be measured quantitatively and the inputs can be varied in terms of their values by the user.
    b. The inputs to the rendering process are independent of one another, but the outputs are dependent on the inputs.
    c. The inputs and outputs are related via mathematical expressions such as dynamic polynomial equations and/or state space equations.

2. A method for controlling user-defined parameters pertaining to the input(s) and output(s) described to Claim (1) of the rendering system whereby:
    a. The correlation described in Claim (1) is represented as a system model and is used to determine the parameters in the controller design.
    b. The controller may be designed by using model-based control design approaches as well as by using model-free approaches.

3. A method for establishing the communication channel between the rendering process and the controller, whereby the following schemes may be implemented:
    a. The control system is implemented in a single computer device/hardware as the rendering process.
    b. The control system is implemented over a network of computer devices/ hardware via a communication channel.

4. A system for controlling real-time computer graphics rendering whereby:
    a. The rendering process is able to meet user defined objectives without human intervention
    b. The rendering process continues to run "online" without the need to stop or any manual effort to work on it "off-line"
    c. The performance objectives are met consistently over a period of time and are sustainable.

5. A method for designing the controller for the rendering system whereby:
    a. A gain scheduling PID controller is used to control a large operating range by cascading several PID controllers
    b. Each PID controller's gain parameters are derived using rendering system models obtained from empirical data via the system identification methodology
    c. A fuzzy controller is used to control the rendering system without the need for any predefined system model of the rendering process
    d. The fuzzy controller's membership functions and rule base are derived from intuitive understanding of the relationship between the input and output of the rendering system.

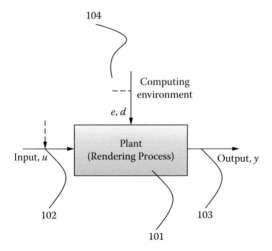

FIGURE B.1   System model of open-loop rendering process.

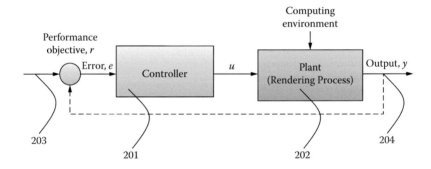

FIGURE B.2   Closed-loop control system with feedback.

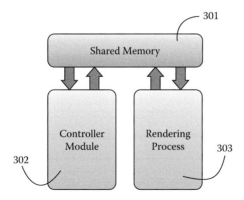

FIGURE B.3   Deployment in single computer device.

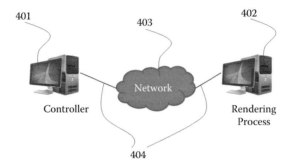

FIGURE B.4   Deployment in distributed computer environment.

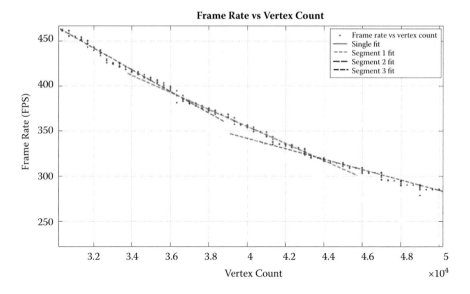

FIGURE B.5   Plot of steady-state values of input and output data.

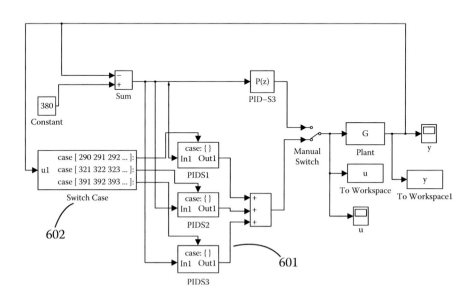

FIGURE B.6   Control system.

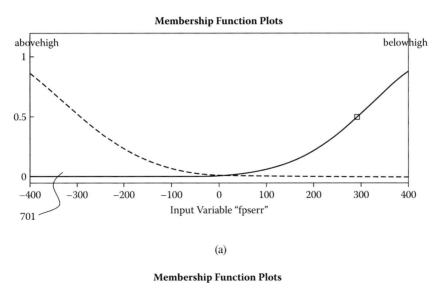

(a)

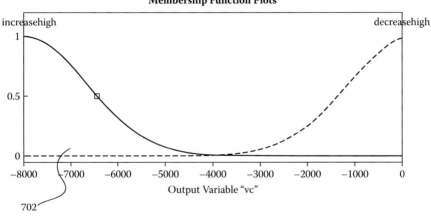

(b)

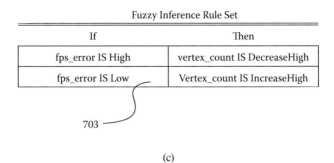

(c)

FIGURE B.7   Relationship of input and output of rendering system.

Annex C: Neural PID Control System Code

```
%Single Neural Adaptive PID Controller Code

clear all;
close all;

%Initialize key variables
% Neuron states
x = [0,0,0]';
xP = 300.001;
xI = 0.85;
xD = 0.25;

% Neuron weights
wkp_1 = 0.50;
wki_1 = 0.50;
wkd_1 = 0.50;

error_1 = 0;
error_2 = 0;

refValue = 356;

y_1 = 0;y_2 = 0;y_3 = 0;
u_1 = 0;u_2 = 0;u_3 = 0;

%load n4s1
load model

%convert idss model to ss
H = ss(n4s1);

%take "measured" channel
plant = tf(H(1) + H(2));
ts = H.Ts;

%get direct form coefficients to use `direct form 1` in loop
b = plant.num{1};
a = plant.den{1};

%%%%%%%%%%%%%%%%%%% First run executes based on SNPID %%%%%%%%%%%%%%%%%%
for k = 1:15000
    time(k) = k*ts;

    if k< = 5000
    % Reference
        rin(k) = 356.0;
```

```
   elseif k< = 10000
      rin(k) = 200;
   else
      rin(k) = 50;
   end


   try
      u(k) = u_1+K*w*x; %Control law
   catch
      u(k) = 0;
   end
   % Plant model
   yout(k) = b(1) * u(k) + b(2) * u_1 + b(3) * u_2 + b(4) * u_3 - a(2)
* y_1 - a(3) * y_2 - a(4) * y_3;
   % Error
   error(k) = rin(k)-yout(k);

%Adjusting Weight Value by hebb learning algorithm
M = 4;
if M = =1       %No Supervised Heb learning algorithm
   wkp(k) = wkp_1+xiteP*u_1*x(1); %P
   wki(k) = wki_1+xiteI*u_1*x(2); %I
   wkd(k) = wkd_1+xiteD*u_1*x(3); %D
   K = 0.06;
elseif M = =2 %Supervised Delta learning algorithm
   wkp(k) = wkp_1+xiteP*error(k)*u_1; %P
   wki(k) = wki_1+xiteI*error(k)*u_1; %I
   wkd(k) = wkd_1+xiteD*error(k)*u_1; %D
   K = 0.12;
elseif M = =3  %Supervised Heb learning algorithm
   wkp(k) = wkp_1+xiteP*error(k)*u_1*x(1); %P
   wki(k) = wki_1+xiteI*error(k)*u_1*x(2); %I
   wkd(k) = wkd_1+xiteD*error(k)*u_1*x(3); %D
   K = 0.12;
elseif M = =4  %Improved Heb learning algorithm
   wkp(k) = wkp_1+xiteP*error(k)*u_1*(2*error(k)-error_1);
   wki(k) = wki_1+xiteI*error(k)*u_1*(2*error(k)-error_1);
   wkd(k) = wkd_1+xiteD*error(k)*u_1*(2*error(k)-error_1);
   K = 0.12;
end

   x(1) = error(k)-error_1;             %P
   x(2) = error(k);                     %I
   x(3) = error(k)-2*error_1+error_2;   %D

   wadd(k) = abs(wkp(k))+abs(wki(k))+abs(wkd(k));
   w11(k) = wkp(k)/wadd(k);
   w22(k) = wki(k)/wadd(k);
   w33(k) = wkd(k)/wadd(k);
   w = [w11(k),w22(k),w33(k)];

error_2 = error_1;
error_1 = error(k);

u_3 = u_2;u_2 = u_1;u_1 = u(k);
```

```
y_3 = y_2;y_2 = y_1;y_1 = yout(k);

wkp_1 = wkp(k);
wkd_1 = wkd(k);
wki_1 = wki(k);
end

%%%%%%%%%%%%%%%%%%%%%%%%%%%%%%%%%%%%%%%%%%%%%%%%%%%%%%%%%%%%%%%%%%%%%%%%%%%%
%%%%%%%%% Second run executes based on Matlab's built-in PID %%%%%%%%%
load model
%convert idss model to ss
H = ss(n4s1);
%take "measured" channel
H = H(1);
ts = H.Ts;
%train pid
C = pidtune(H, 'pi');

plantWithPid = feedback(series(C, H), 1);
% rin->(+)->[C]-> [H]—->yout
% ^   |
% |————-|

%constant rin
%pid_rin = refValue * ones(1, length(time));
pid_rin = rin;
pid_yout = lsim(plantWithPid, pid_rin, time);
pid_error = pid_rin' - pid_yout;

%to find pid's output send rin-yout to pid
%rin-yout' -> [pid] -> u
pid_u = lsim(C, pid_rin-pid_yout', time);

%test original system with pid signal
%pid_yout2 = lsim(n4s1, pid_u, time);

%Output graphs
figure 1);
hold on;
plot(time,rin,'b',time,yout,'r');
plot(time,pid_rin,'k',time,pid_yout,'k');
xlabel('Frame');ylabel('rin,yout');
legend('rin','yout', 'pid yout');

%Error graphs
figure 2);
plot(time,error,'r', time, pid_error, 'b');
xlabel('Frame');ylabel('error');
legend('error', 'pid error');

%SNPID control input
%figure 3);
%plot(time,u,'r');
%xlabel('Frame');ylabel('Control Input');
%legend('u', 'pid u');
```

```
%PID control input
%figure 4);
%plot(time,pid_u,'g');
%xlabel('Frame');ylabel('Control Input');
%legend('u', 'pid u');

figure 5);
subplot(2,1,1);
plot(time,u,'r');
xlabel('Frame');ylabel('Control Input');
subplot(2,1,2);
plot(time,pid_u,'g');
xlabel('Frame');ylabel('Control Input');
```

References

1. L. Ljung, *System Identifiation*, 2nd Edition, Prentice Hall, 1999, ISBN 978-0136566953.
2. T. F. Abdelzaher, J. A. Stankovic, C. Lu, R. Zhang, and Y. Lu, "Feedback Performance Control in Software Services," *IEEE Control Systems Magazine*, vol. 23(3), June 2003.
3. S. Abdelwahed, N. Kandasamy, S. Neema, "Online Control for Self-Management in Computing Systems," *Real-Time and Embedded Technology and Applications Symposium*, IEEE, pp. 368, 2004.
4. B. Li, K. Nahrstedt, "A Control-based Middleware Framework for Quality of Service Adaptations", *IEEE Journal of Selected Areas in Communication, Special Issue on Service Enabling Platforms*, vol. 17(9), pp. 1632–1650, Sept. 1999.
5. J. Hellerstein, S. Singhal, and Q. Wang, "Research challenges in control engineering of computing systems", *IEEE Trans. on Network and Server Management*, pp. 206–211, Dec. 2009.
6. J. Hellerstein, Y. Diao, S. Parekh and D. Tilbury, *Feedback Control of Computing Systems*, Wiley, 2004, ISBN-13: 978-0471266372.
7. T. Abdelzaher, Y. Diao, J. L. Hellerstein, C. Lu and X. Zhu, "Introduction to Control Theory and its Application to Computing Systems", *Performance Modeling and Engineering*, Springer, pp. 185–215, 2008.
8. Y. Lu, T. Abdelzaher, C. Lu, L. Sha and X. Liu, "Feedback Control with Queueing-Theoretic Prediction for Relative Delay Guarantees in Web Servers," *Proceedings of the 9th IEEE Real-Time and Embedded Technology and Applications Symposium*, Washington DC, pp. 208–217, 2003.
9. C. Karamanolis, M. Karlsson, and X. Zhu, "Designing controllable computer systems", *Proceedings of the 10th conference on Hot Topics in Operating Systems*, Vol. 10, Berkeley, CA, USA, 2005.
10. X. Li and H. Shen, "Adaptive Volume Rendering using Fuzzy Logic", *Proceedings of Joint Eurographics-IEEE TCVG Symposium on Visualization*, Springer-Verlag, pp. 253–262, 2001.
11. Y. Kirihata, J. Leigh, C. Xiong and T. Murata, "A Sort-Last Rendering System Over an Optical Backplane," *Proceedings of the 10th International Conference on Information Systems Analysis and Synthesis*, vol. 1, pp. 42–47, Orlando, Florida July, 2004.
12. D. Cohen-Or et al, "A Survey of Visibility for Walkthrough Applications," *IEEE Transactions on Visualization and Computer Graphics*, pp. 412–431, July, 2003.
13. E. Haines, "An Introductory Tour of Interactive Rendering", *IEEE Computer Graphics and Applications*, vol. 26(1), pp. 76–87, Jan./Feb. 2006.
14. T. Akenine-Moller, E. Haines and N. Hoffman, *Real-time Rendering*, 3rd Edition, A.K. Peters, 1045 pages, 2008, ISBN 1568814240.
15. Microsoft Direct3D 11, http://msdn.microsoft.com/en-us/library/ff476340(VS.85).aspx
16. S. Kyöstilä, K.J. Kangas, and K. Pulli, "Tracy: A Debugger and System Analyzer For Cross-Platform Graphics Development," *Proceedings of the 23rd ACM SIGGRAPH/ EUROGRAPHICS Symposium on Graphics Hardware*, pp. 1–11, 2008.
17. J.R. Monfort and M. Grossman, "Scaling of 3D game engine workloads on modern multi-GPU systems," *Conference on High Performance Graphics*, pp. 37–46, 2009.
18. N. Tack, F. Morán, G. Lafruit, and R. Lauwereins, "3D Graphics Rendering Time Modeling and Control for Mobile Terminals," *Proceedings of the Ninth International Conference on 3D Web Technology*, pp. 109–117, 2004.
19. M. Wimmer, and P. Wonka, "Rendering time estimation for real-time rendering," *14th Eurographics Workshop on Rendering*, vol. 44. pp. 118–129, 2003.

20. J.T. Klosowski, and C.T. Silva, "Rendering on a Budget: A Framework for Time-Critical Rendering," *Proceedings of the Conference on Visualization*, pp. 115–122, IEEE Computer Society Press, 1999.

21. I. Wald, A. Dietrich, and P. Slusallek, "An Interactive Out-Of-Core Rendering Framework for Visualizing Massively Complex Models," *ACM SIGGRAPH,* Courses notes, 2005.

22. A. Lakhia, "Efficient Interactive Rendering of Detailed Models with Hierarchical Levels of Detail", *2nd International Symposium on 3D Data Processing, Visualization and Transmission*, pp. 275–282, Sept. 2004.

23. T.A. Funkhouser and C.H. S´Equin, "Adaptive Display Algorithm for Interactive Frame Rates during Visualization of Complex Virtual Environments," *Proc. 20th Annual Conference on Computer Graphics and Interactive Techniques (ACM SIGGRAPH '93)*, pp. 247–254, 1993.

24. E. Gobbetti, and E. Bouvier, "Time-critical Multiresolution Scene Rendering," *Proc. Conference on Visualization,* pp. 123–130, 1999.

25. Microsoft Xbox 360 Technical Specifications, http://support.xbox.com/support/en/us/xbox360/hardware/specifications/consolespecifications.aspx

26. Sony PlayStation Specifications, http://playstation.about.com/od/ps3/a/PS3SpecsDetails_3.htm

27. N. Tack, F. Morán, G. Lafruit, and R. Lauwereins, "3D Graphics Rendering Time Modeling and Control for Mobile Terminals," *Proceedings of the Ninth International Conference on 3D Web Technology*, pp. 109–117, 2004.

28. D. Luebke, B. Watson, J.D. Cohen, M. Reddy, and A, Varshney, *Level of Detail for 3D Graphics*, Elsevier Science, 2002, ISBN-13: 978-0123991812.

29. R. Dumont, F. Pellacini and J.A. Ferwerda, "Perceptually-Driven Decision Theory for Interactive Realistic Rendering," *ACM Transactions on Graphics*, vol. 22(2), pp. 152–181, Apr. 2003.

30. NVIDIA, DirectX 11 Tessellation—What It Is and Why It Matters, http://www.nvidia.com/object/tessellation.html

31. MATLAB, The MathWorks Inc., http://www.mathworks.com

32. M. Claypool and K. Claypool, "Perspectives, Frame Rates and Resolutions: it's all in the Game," *Proc. 4th International Conference on Foundations of Digital Games*, pp. 42–49, 2009.

33. K. Claypool and M. Claypool, "On Frame Rate and Player Performance in First Person Shooter Games," Proc. Multimedia Systems, vol. 13(1), pp. 3–17, 2007.

34. P. Yuan, M. Green, and R.W. Lau, "A Framework for Performance Evaluation of Real-Time Rendering Algorithms in Virtual Reality," *Proc. ACM Symposium on Virtual Reality Software and Technology*, pp. 51–58, 1997.

35. H.M. Sun, Y.C. Lin, and L. Shu, "The Impact of Varying Frame Rates and Bit Rates on Perceived Quality of Low/High Motion Sequences with Smooth/Complex Texture," *Proc. Multimedia Systems*, vol. 14(1), pp. 1–13, 2007.

36. B. Watson, V. Spaulding, N. Walker and W. Ribarsky, "Evaluation of the Effects of Frame Time Variation on VR Task Performance," *Proc. IEEE Virtual Reality Annual International Symposium*, p. 38, 1997.

37. R. Hawkes, S. Rushton, and M. Smyth, "Update Rates and Fidelity in Virtual Environments," *Virtual Reality: Research, Applications and Design*, vol 1(2), pp. 99–108, 1995.

38. B. Hook and A. Bigos, 3D Acceleration Demystified, Part II: The Benchmarks, http://www.gamasutra.com/features/19970601/3d_acceleration_demystified.htm.

39. W. Mcculloch and W. Pitts, "A Logical Calculus of Ideas Immanent in Nervous Activity", *Bulletin of Mathematical Biophysics*, pp. 115–133, 1943.

40. D.E. Rumelhart, G.E. Hinton, R.J. Williams, "Learning Internal Representations by Error Propagation", vol. 1, MIT Press, Cambridge, MA, 1986.

41. D.W. Marquardt, "An Algorithm for Leastsquares Estimation of Nonlinear Parameters", *Journal of the Society for Industrial and Applied Mathematics* vol. 11(1), pp. 431–444, 1963.

42. A. Waibel, T. Hanazawa, G. Hilton, K. Shikano, and K. J. Lang, "Phoneme recognition using time-delay neural networks," *IEEE Transactions on Acoustics, Speech, and Signal Processing*, vol. 37, pp. 328–339, 1989.

43. G. Chen, *Introduction to Fuzzy Sets, Fuzzy Logic, and Fuzzy Control Systems*, CRC Press, 328 pages, 2000, ISBN 9780849316586.

44. Microsoft DirectX—http://msdn.microsoft.com/en-US/directx/

45. NVIDIA PerfKit, http://developer.nvidia.com/object/nvperfkit_home.html

46. H. Demuth, M. Beale, *Neural Network Toolbox for use with MATLAB—User's Guide*. The Mathworks, USA, 1993.

47. J.R. Jang, "ANFIS: Adaptive-Network-based Fuzzy Inference Systems," *IEEE Trans. on Systems, Man, and Cybernetics*, vol. 23, pp. 665–685, May 1993.

48. W. Barfield and C. Hendrix, "The effect of update rate on the sense of presence within virtual environments", *Virtual Reality,* vol. 1(1), pp. 3–15, 1993.

49. J. Chen and J. Thropp, "Review of low frame rate effects on human performance", *IEEE Transactions on Systems, Man and Cybernetics* vol. 37(6) pp. 1063–1076, Nov. 2007.

50. S. J. Qin and T. A. Badgwell, "A survey of industrial model predictive control technology", *Control Engineering Practice*, vol. 11(7), pp. 733–764. Jul. 2003.

51. B. Wittenmark, "A survey of adaptive control applications", *Dynamic Modeling Control Applications for Industry Workshop*, IEEE Industry Applications Society, pp. 32–36, 1997.

52. M. Kokar, K. Baclawski and Y. Eracar, "Control Theory-Based Foundations of Self-Controlling Software", *IEEE Intelligent Systems and their Applications*, vol. 14(3), pp. 37–45, 1999.

53. K. J. Astrom and B. Wittenmark, *Adaptive Control*, Addison-Wesley Longman Publishing Co., Inc., Boston, MA, USA, 1994.

54. J. L. Hellerstein, Y. Diao, S. Parekh and D. Tilbury, *Feedback Control of Computing Systems*. John Wiley and Sons., 2004, ISBN-13: 978-0471266372.

55. K. J. Astrom and T. Hagglund, *PID Controllers: Theory, Design and Tuning*, International Society for Measurement and Control, 343 pages, 1995, ISBN 978-1556175169.

56. Automated Tuning of Simulink PID Controller, http://www.mathworks.com/help/toolbox/slcontrol/gs/bs1qetr.html

57. R. J. Craddock, K Warwick, "The use of state space control theory for analyzing feed-forward neural networks", *Dealing with Complexity: a Neural Network Approach*, Springer, 1998.

58. H. Stone, "Approximation of curves by line segments," *Math. Comp.*, vol. 15, pp. 40–47, 1961,

59. R. Bellman, "On the approximation of curves by line segments using dynamic programming". *ACM Comm.* vol. 4(6), pp. 284, June 1961.

60. W. S. Chan and F. Chin. 1992, "Approximation of Polygonal Curves with Minimum Number of Line Segments", In *Proceedings of the Third International Symposium on Algorithms and Computation*, Springer-Verlag, London, UK, pp. 378–387, 1992.

61. L. McMillan and G. Bishop, "Plenoptic modeling: an image-based rendering system," In *Proceedings of the 22nd annual conference on Computer graphics and interactive techniques* (SIGGRAPH '95), ACM, New York, NY, USA, pp. 39–46, 1995.

62. D. Seo and I. Jung, "Network-adaptive autonomic transcoding algorithm for seamless streaming media service of mobile clients," *Multimedia Tools Appl.* vol. 51(3), pp. 897–912, Feb. 2011.

63. C. Li, G. Feng, W. Li, T. Gu, S. Lu, and D. Chen, "A resource-adaptive transcoding proxy caching strategy," In Proceedings of the *8th Asia-Pacific Web conference on Frontiers of WWW Research and Development* (APWeb'06), Springer-Verlag, Berlin, Heidelberg, pp. 556–567, 2006.

64. H. Fang, X. Yu, "Design and Simulation of Neuron PID Controller," *International Conference on Information Technology, Computer Engineering and Management Sciences* (ICM), 2011, vol. 1, pp. 80–82, 24–25 Sept. 2011

65. A. Niels et al., "An application framework for adaptive distributed simulation and 3D rendering services", *Proceedings of the 11th ACM SIGGRAPH International Conference on Virtual-Reality Continuum and its Applications in Industry (VRCAI '12)* ACM, NY, USA, pp. 103–110, 2012.

66. G. Paravati, A. Sanna, F. Lamberti, and L. Ciminiera, "An Adaptive Control System to Deliver Interactive Virtual Environment Content to Handheld Devices" *Mobile Networking Applications*, vol. 16(3), pp. 385–393. Jun. 2011.

67. N.A. Nijdam. S. Han, B. Kevelham, N. Magnenat-Thalmann, "A context-aware adaptive rendering system for user-centric pervasive computing environments," *15th IEEE Mediterranean Electrotechnical Conference*, pp. 790–795, April 2010.

68. L. Hu, P.V. Sander and H. Hoppe, "Parallel View-Dependent Level-of-Detail Control," *IEEE Transactions on Visualization and Computer Graphics*, vol. 16, no. 5, pp. 718, 728, Sept.–Oct. 2010.

69. D. Scherzer, L. Yang, and O. Mattausch. "Exploiting temporal coherence in real-time rendering", *ACM SIGGRAPH ASIA 2010 Courses* (SA '10). ACM, NY, USA, Article 24, 26 pages, 2010.

70. Y. Huai, X. Zeng, P. Yu, J. Li, "Real-time rendering of large-scale tree scene," *4th International Conference on Computer Science and Education*, 2009. ICCSE '09, pp. 748–752, 25–28 July 2009.

71. Z. Zheng, E. Prakash and T.K.Y. Chan, "Interactive View-Dependent Rendering over Networks," *IEEE Transactions on Visualization and Computer Graphics*, vol. 14, no. 3, pp. 576, 589, May–June 2008.

72. R.W.N Pazzi, A. Boukerche, T. Huang, "Implementation, Measurement, and Analysis of an Image-Based Virtual Environment Streaming Protocol for Wireless Mobile Devices," *IEEE Transactions on Instrumentation and Measurement*, vol. 57, no. 9, pp. 1894, 1907, Sept. 2008.

73. Y. Gu and S. Chakraborty, "A Hybrid DVS Scheme for Interactive 3D Games," *Real-Time and Embedded Technology and Applications Symposium*, 2008. RTAS '08. IEEE, pp. 3–12, 22–24 April 2008.

74. A. Domingo et al., "Continuous LODs and Adaptive Frame-Rate Control for Spherical Light Fields," *Geometric Modeling and Imaging*, 2007. GMAI '07, vol., no., pp. 73, 78, 4–6 July 2007.

75. J. Kuo, G. R. Bredthauer, J. B. Castellucci, O.T. Von Ramm, "Interactive volume rendering of real-time three-dimensional ultrasound images," *IEEE Transactions on Ultrasonics, Ferroelectrics and Frequency Control*, vol. 54, no. 2, pp. 313–318, Feb. 2007.

76. Z. Zheng, T.K.Y. Chan, and P. Edmond, "Rendering of large 3D models for online entertainment", In Proceedings of the *2006 international conference on Game research and development* (CyberGames '06), Murdoch University, Australia, pp. 163–170. 2006.

77. S. Jeschke, M. Wimmer, H. Schumann, and W. Purgathofer, "Automatic impostor placement for guaranteed frame rates and low memory requirements," In Proceedings of the *2005 symposium on Interactive 3D graphics and games* (I3D '05). ACM, NY, USA, pp. 103–110, 2005.

78. J. Pouderoux and J-E. Marvie, "Adaptive streaming and rendering of large terrains using strip masks," In Proceedings of the *3rd international conference on Computer graphics and interactive techniques in Australasia and South East Asia* (GRAPHITE '05). ACM, NY, USA, pp. 299–306, 2005.

79. X. Li and Q. He, "Frame rate control in distributed game engine," In Proceedings of the *4th international conference on Entertainment Computing* (ICEC'05), Springer-Verlag, Berlin, Heidelberg, pp. 76–87, 2005.

80. M. Wan, N. Zhang and H. Qu, A.E. Kaufman, "Interactive stereoscopic rendering of volumetric environments," IEEE Transactions on *Visualization and Computer Graphics*, vol. 10, no. 1, pp. 15–28, Jan.–Feb 2004.

81. S-E. Yoon, B. Salomon, R. Gayle, and D. Manocha. "Quick-VDR: Interactive View-Dependent Rendering of Massive Models," In Proceedings of the *conference on Visualization '04 (VIS '04),* IEEE Computer Society, Washington, DC, USA, pp. 131–138, 2004.

82. N. Tack, F. Morn, G. Lafruit, and R. Lauwereins, "3D graphics rendering time modeling and control for mobile terminals," In Proceedings of the *ninth international conference on 3D Web technology* (Web3D '04). ACM, New York, NY, USA, pp. 109–117, 2004.

83. R. Dumont, F. Pellacini, and J. A. Ferwerda, "Perceptually-driven decision theory for interactive realistic rendering" *ACM Trans. Graph.* vol. 22(2), pp. 152–181, April 2003.

84. X. Li and H-W. Shen, "Time-critical multiresolution volume rendering using 3D texture mapping hardware," Proceedings IEEE/ACM SIGGRAPH Symposium on *Volume Visualization and Graphics*, pp. 29, 36, Oct. 2002.

85. M. Grabner, "Smooth high-quality interactive visualization," Spring Conference on *Computer Graphics*, pp. 87, 94, 2001.

86. W-S. Lin; R. Lau, W.H. Kai Hwang; X. Lin; P.Y.S Cheung, "Adaptive parallel rendering on multiprocessors and workstation clusters," Parallel and Distributed Systems, IEEE Transactions on, vol. 12, no. 3, pp. 241, 258, Mar 2001

87. H. Qu, M. Wan, J. Qin, and A. Kaufman, "Image based rendering with stable frame rates," In Proceedings of the *conference on Visualization* '00 (VIS '00). IEEE Computer Society Press, Los Alamitos, CA, USA, pp. 251–258, 2000.

88. E. Gobbetti and E. Bouvier, "Time-critical multiresolution scene rendering," In Proceedings of the *conference on Visualization '99*: celebrating ten years (VIS '99). IEEE Computer Society Press, Los Alamitos, CA, USA, pp. 123–130, 1999.

89. J. T. Klosowski and C. T. Silva, "Rendering on a budget: a framework for time-critical rendering," In Proceedings of the *conference on Visualization* '99: celebrating ten years (VIS '99). IEEE Computer Society Press, Los Alamitos, CA, USA, 115–122. 1999

90. P. Ebbesmeyer, "Textured virtual walls achieving interactive frame rates during walk-throughs of complex indoor environments," Proceedings IEEE *Virtual Reality Annual International Symposium*, pp. 220–227, 1998.

91. J. Sato, K. Hashimoto, Y. Shibata, "Dynamic rate control methods for continuous media transmission," Proceedings Twelfth International Conference on *Information Networking*, (ICOIN-12), pp. 110, 115, 21–23 Jan 1998.

92. M. Zockler, D. Stalling and H.-C. Hege, "Interactive visualization of 3D-vector fields using illuminated stream lines," *Visualization '96*. Proceedings, pp. 107–113, Nov. 1 1996.

93. S. Belblidia, J.-P. Perrin and J.C. Paul, "Generating various levels of detail of architectural objects for image-quality and frame-rate control rendering," *Computer Graphics International, 1996. Proceedings*, pp. 84–89, 24–28 Jun 1996.

94. S. Bryson and S. Johan, S, "Time management, simultaneity and time-critical computation in interactive unsteady visualization environments," *Visualization '96*, pp. 255–261, Nov. 1 1996.

95. T.L. Kunii and S. Nishimura, "Parallel polygon rendering on the graphics computer VC-1," *Proceedings First Aizu International Symposium* on *Parallel Algorithms/Architecture Synthesis*, pp. 2–9, 15–17 Mar 1995.

96. T. Tamada, Y. Nakamura, S. Takeda, "An efficient 3D object management and interactive walkthrough for the 3D facility management system," *20th International Conference on Control and Instrumentation*, 1994. vol. 3, pp. 1937–1941, Sep 1994.

97. A. Dayal, C. Woolley, B.A. Watson and D. Luebke, "Adaptive frameless rendering", *Proc. Eurographics Symposium on Rendering*, pp. 265–275, 2005.

Publications and Achievements

PATENT APPLICATION

Method and System for Adaptive Control of Real-time Computer Graphics Rendering, US Patent Office dated 13 July 2011, US61/507,486. (Supported by Nanyang Enterprise and Innovation Office)

BOOK

G. Wong and J. Wang, *Computer Graphics with Control Engineering*, to be published by Taylor and Francis, CRC Press. ISBN 978-1466583597.

BOOK CHAPTERS

G. Wong and J. Wang, "Intelligent Load Control Shader", *ShaderX7*, Charles River Media/Thomson, 2009, ISBN 978-1584505983, pp. 627-633, March 2009.

G. Wong and J. Wang, "A Fuzzy Control Approach to Managing Scene Complexity", Charles River Media/Thomson, *Games Programming Gems 6*, 2006, ISBN 1584504501, pp. 305–314.

CONFERENCE PAPERS

G. Wong and J. Wang, "Control of Interactive Computer Graphics Rendering Process", *9th IEEE International Conference on Control & Automation* (IEEE ICCA'11), December 2011, Chile.

G. Wong and J. Wang, "Dynamics of 3D Polygonal Rendering", *IEEE R8 International Conference on Computational Technologies in Electrical and Electronics Engineering*, SIBIRCON 2010, July 11–15, 2010, Irkutsk Listvyanka, Russia.

G. Wong and J. Wang, "Green Graphics: Feedback Control for Energy Efficient Rendering", *International Conference on Computer Graphics and Interactive Techniques* (ACM SIGGRAPH Asia 2008), ISBN 978-1-60558-388-4, Singapore, 2008

G. Wong and J. Wang, "Control Theory based Real-time Rendering", *Proceedings of the 7th International Conference on Virtual Reality Continuum and its Applications in Industry*, VRCAI 2008, Singapore, December 8–9, 2008, ISBN 978-1-60558-335-8.

G. Wong and J. Wang, "Interactive Rendering of Dynamic Environment using PID Control", *International Conference on Computer Graphics and Interactive Techniques* (ACM SIGGRAPH 2007), ISBN 978-1-59593-648-6.

G. Wong and J. Wang, "Modeling Real-time Rendering", *EUROGRAPHICS Conference 2006*, Vienna, Austria, ISSN 1017-4656, pp. 89–93.

G. Wong and J. Wang, "Efficient Level-of-Detail Management using Fuzzy Logic", *International Conference on Computer Graphics and Interactive Techniques* (ACM SIGGRAPH 2005), ISBN 1-59593-100-7.

ACHIEVEMENTS

Academic Research Grant Tier 1 (Grant number: RG26/09), NTU/Ministry of Education
     Singapore.
Research fund support by Defense Science Organization, Singapore under grant DSOCL06184.
Microsoft Research Student Competition, 2005, Semi-finalist.

Index

A

Adaptive neuro-fuzzy inference system (ANFIS), 57–60, 90–92
    experiments, 57–60, 92–97
Adaptive tracking, 112–117
Artificial neural networks (ANNs), *See* Neural networks
Auto-regressive exogenous (ARX) model, 12, 25–27, 45, 48*t*, 156
Auto-regressive moving average (ARMA) model, 12

B

Black-box modelling, 10, 11, 12, 23
Bottleneck identification, 22
Bounded-input-bounded output (BIBO), 70
Box-Jenkins model, 12

C

Camera view coordinate system transformation, 6
Clipping, 7
Colour of pixels, 7, 8
Computer-aided design (CAD) and -manufacturing (CAM) applications, 2, 104, 106
Computer game applications, 104
Computer vision algorithms, 9
Control systems, 3, 67–68, 92–94
    applications, 104–105
    architectures, 68–69, 99–101
    code for neural PID control system, 167–170
    consequences of ineffective systems, 1
    control-centric rendering time control, 21
    data preprocessing, 75
    delay impacts, 97–98
    economic and productivity impacts, 106–107
    experiments and results, 29–41, 81–82, 83–86, 92–94
        sample applications and code, 121–152
    extension, 105
    future work, 118–119
    fuzzy logic, 17, 53–55, *See also* Fuzzy control
    gain scheduling, 78, 82, 85–87, 156–159
    heuristics, 21–22
    implementing for rendering process, 72–75
    key objectives, 2
    literature review, 14–17
    model-less approach, 89, *See also* Fuzzy control
    modular system design, 69
    neural control, 79–81
    patent for method and system, 153–165
        fuzzy (model-less) control, 159–160
        PID gain scheduling, 156–159
        preliminary claims, 161
    performance, 70–71, 103, *See also* Performance, comparison issues
        algorithm complexity, 104
        data integrity, 103
        data structures and handling, 103–104
        plant-controller communication latency, 103
    PID, *See* Proportional, integral, derivative (PID) controllers
    rendering process model, 67–68
    software design, 101–102
    tuning, 71, 74, 79
    units of measurement, 68
    Windows operating system limitations, 72–74
Coordinate system transformations, 5–7
Culling, 6, 117

D

Data collection, 11–12, 14
Data-driven modelling methodology, 3, 23–25, 41, *See also* System identification methodology
    comparison with other estimation techniques, 41–42
Data integrity issues, 103
Data preprocessing in PID control systems, 75
Data structures and handling, 103–104
Debugging, 22
Defuzzification, 54, 55, 160
DirectX, 24, 56, 121
    performance counters, 23–24
    tessellation (subdivision), 27–28
    test application, 44
Display stuttering, 112
Display window coordinates transformation, 7
Distributed time delay neural network (DTDNN), 52
    linearisation, 64–65

E

Economic and productivity impacts, 106–107